PALGRAVE STUDI
SCIENCE AN

M000307381

James Rodger Fleming (Colby College) and Roger D. Launius (National Air and Space Museum), Series Editors

This series presents original, high-quality, and accessible works at the cutting edge of scholarship within the history of science and technology. Books in the series aim to disseminate new knowledge and new perspectives about the history of science and technology, enhance and extend education, foster public understanding, and enrich cultural life. Collectively, these books will break down conventional lines of demarcation by incorporating historical perspectives into issues of current and ongoing concern, offering international and global perspectives on a variety of issues, and bridging the gap between historians and practicing scientists. In this way they advance scholarly conversation within and across traditional disciplines but also to help define new areas of intellectual endeavor.

Published by Palgrave Macmillan:

*Continental Defense in the Eisenhower Era: Nuclear Antiaircraft Arms and the Cold War*
By Christopher J. Bright

*Confronting the Climate: British Airs and the Making of Environmental Medicine*
By Vladimir Janković

*Globalizing Polar Science: Reconsidering the International Polar and Geophysical Years*
Edited by Roger D. Launius, James Rodger Fleming, and David H. DeVorkin

*Eugenics and the Nature-Nurture Debate in the Twentieth Century*
By Aaron Gillette

*John F. Kennedy and the Race to the Moon*
By John M. Logsdon

*A Vision of Modern Science: John Tyndall and the Role of the Scientist in Victorian Culture*
By Ursula DeYoung

*Searching for Sasquatch: Crackpots, Eggheads, and Cryptozoology*
By Brian Regal

*Inventing the American Astronaut*
By Matthew H. Hersch

*The Nuclear Age in Popular Media: A Transnational History*
Edited by Dick van Lente

*Exploring the Solar System: The History and Science of Planetary Exploration*
Edited by Roger D. Launius

*The Sociable Sciences: Darwin and His Contemporaries in Chile*
By Patience A. Schell

*The First Atomic Age: Scientists, Radiations, and the American Public, 1895–1945*
By Matthew Lavine

*NASA in the World: Fifty Years of International Collaboration in Space*
By John Krige, Angelina Long Callahan, and Ashok Maharaj

# Searching for Sasquatch

## Crackpots, Eggheads, and Cryptozoology

Brian Regal

First published in hardcover in 2011 by PALGRAVE MACMILLAN® in the United States—a division of St. Martin's Press LLC, 175 Fifth Avenue, New York, NY 10010.

Where this book is distributed in the UK, Europe and the rest of the world, this is by Palgrave Macmillan, a division of Macmillan Publishers Limited, registered in England, company number 785998, of Houndmills, Basingstoke, Hampshire RG21 6XS.

Palgrave Macmillan is the global academic imprint of the above companies and has companies and representatives throughout the world.

Palgrave® and Macmillan® are registered trademarks in the United States, the United Kingdom, Europe and other countries.

ISBN: 978–1–137–34943–9

The Library of Congress has cataloged the hardcover edition as follows:

Regal, Brian.
    Searching for sasquatch : crackpots, eggheads, and cryptozoology / by Brian Regal.
        p. cm.—(Palgrave studies in the history of science and technology)
        Includes bibliographical references.
        ISBN 978–0–230–11147–9 (alk. paper)
        1. Sasquatch. 2. Monsters. 3. Cryptozoology. 4. Tracking and trailing. I. Title.

QL89.2.S2R44 2011
001.944—dc22                                    2010036009

A catalogue record of the book is available from the British Library.

Design by Newgen Imaging Systems (P) Ltd., Chennai, India.

First PALGRAVE MACMILLAN paperback edition: July 2013

10 9 8 7 6 5 4 3 2 1

Transferred to Digital Printing in 2013

# Contents

# A Note About the Cover Image

This image of a smiling, seeming Sasquatch is from the cover of *Bickerstaff's Boston Almanac* for 1785. The image is actually a crude copy of the famous illustration of a chimpanzee from Edward Tyson's pioneering work of primate morphology *Anatomy of the Pygmie* (1699). It is likely the first printed image of a primate published in North America, and may have contributed to popular conceptions about what a Sasquatch looks like.

# Acknowledgments

I would like to thank the following people for giving me their time and comments most graciously despite the misgivings of some in the monster cognoscenti about how I was going to treat the field. Robert Ackerman, John Bodley, Loren Coleman, Joe Davis, Jonathan Downes, Gregory Forth, Richard Freeman, John Schoenherr, Rick Sprague, Alice Walters, and Milford Wolpoff supplied files and shared insights into monster hunting. Some gave me fascinating anecdotes recorded nowhere before now, but in their personal experiences. Thanks to Eric Altman, Raymond Rosa, Bob Schmalzbach, and all the Bigfooters who took the time to fill out my surveys. Archivists and librarians across North America and the United Kingdom, particularly at the National Anthropological Archive of the Smithsonian Institution, the American Museum of Natural History, the American Philosophical Society, the British Museum, Natural History, the special collections library of University College London, and others showed endless patience in helping me root out obscure papers' collections and primary source materials. Garland Allan, Peter Bowler, and Joe Cain showed me great friendship and this project moral support and encouragement: support that, had it not been given, may have led me to abandon it. I presented a number of the ideas put forward in this book as papers at meetings of the History of Science Society, British Society for the History of Science, and as a guest speaker at the Grant Museum's Darwin Theater Lecture Series, London, between 2007 and 2009. Each time, audience members made thoughtful critiques and suggestions for further study and showed great enthusiasm for the subject and my project. A number of the people listed here, including Alice Wyman, Christopher Bellitto, and Henry Nicholls, read all or part of the work and made suggestions for improvement and clarity (though in the end I take full responsibility for any factual mistakes). Thanks to the blind reviewers for their insightful and useful critiques, which enhanced the manuscript. Thanks to the staff at

Palgrave, including copy editor Stuart A.P. Murray, and to *Palgrave Studies in the History of Science and Technology* series editors Roger Launius of the Smithsonian and James Rodger Fleming of Colby College for their early enthusiasm.

Thanks to Kean University's Department of History Chair Sue-Ellen Gronewold and faculty whose friendship and collegiality helped more than they could know. Thanks to Maria Perez of the faculty development office for so much. The university administration and the Kean Foundation gave me support through a Presidential Research Initiative award for travel and research, and for allowing faculty members avenues for producing research and scholarship and for sharing that research with colleagues and students. This project would not have been possible without it.

Finally, and always, thanks to Lisa Nocks. You know.

# Abbreviations

| | |
|---|---|
| ABSM | Abominable Snowman |
| AMNH | American Museum of Natural History |
| APS | American Philosophical Society |
| BFRO | Bigfoot Research Organization |
| CCNAA | Carleton Coon papers National Anthropological Archive |
| GSK | Grover Sanders Krantz |
| ISC | International Society of Cryptozoology |
| NAA | National Anthropological Archive |
| SMITH | Smithsonian Institution |
| UCL | University College London, Special Collections Library |
| WSU | Washington State University |

# Introduction

# Chasing Monsters

*I know, Watson, that you share my love of all that is bizarre and outside
the conventions and humdrum routine of everyday life.*

Sherlock Holmes, 1891

This story tells of dreams that do not come true. It is a story about
spending one's life pursuing something and never catching it. It is a
story about chasing monsters. These monsters are apish and disturb-
ingly like us. They have stalked the dark parts of the human psyche,
as well as forests, for millennia. If they are real, they are older than we
are, yet they leave behind only footprints and questions. They have
appeared around the world and by various names, including Yeti,
Almasti, Sasquatch, Hibagon, Windigo, Agogwe, Orang-Pendek,
Bigfoot, and many others. Like shadows in the rain, these hard to
explain animals are said to show up in places and in forms conventional wisdom says they should not. Some believe them real animals,
while others scoff at them as so much foolishness. They can be collectively known as manlike monsters, mystery apes, or anomalous
primates. The story, however, concerns more than just the startling
and controversial nature of monsters and monster hunting in the late
twentieth and early twenty-first centuries, but the more important
relationship between the academic scientists and amateur naturalists
who hunt them, and the historiography of unusual scientific evidence.
This discourse exists outside of whether Bigfoot is a biological reality,
a piece of indigenous performance art, or a creature of pop culture.

Situating the life and career of American paleoanthropologist
Gordon "Grover" Sanders Krantz (1931–2002) in the foreground of
the story constitutes an opportune way to study twentieth-century
monster hunting and its place in the history of science. The trials and

tribulations of the emergent field of cryptozoology to find legitimacy and respect as a scientific discipline runs parallel to Krantz's program to establish, not just the reality of Bigfoot, but to establish his own legitimacy by negotiating the complex world of amateur naturalists and academic scientists interested in the Sasquatch question and in whether anomalous primates even rate being a question. As a trained academic, he worked to prove the creature's reality by applying to the problem the techniques of physical anthropology and evolutionary theory: methodologies and theoretical models outside the experience of most of the amateur enthusiasts who have dominated the field of monster studies. He associated, both personally and by reputation, with all the major personalities in both the academic and amateur camps. From 1969 until his death in 2002, Krantz—always known as Grover—played a role in many of the important monster incidents, acting as a linking mechanism between the independent researchers and the academe. His efforts resulted in his being dismissed or ignored by many of his university colleagues, who viewed the Sasquatch as at best a relic of folklore and at worst as a hoax, and Krantz's project to study it as dubious. Krantz also received a negative reaction from some amateur Sasquatch researchers, who threatened and insulted him. At the same time he was respected and liked by others. He claimed his Sasquatch work affected his promotions, funding, and professional status (all things crucial to a successful academic life), yet he devoted his career of more than 30 years to it and never wavered in his belief. His career helps answer the central question of this study: if mainstream scientific thinking held that anomalous primate, manlike monsters did not and could not exist, what motivated so many scientists and academics to pursue them anyway? Why would they associate themselves with the amateurs? Why would they risk their careers? There had to be something to it. Looking beyond the usual "monster theory" of psychological and pop culture explanations leaves only the natural history. It leaves the science. All these academics pursued monsters because they believed the creatures to be real animals.

Reports and legends about hairy humanoids in North America preceded the arrival of Europeans. Beliefs in wild men and kindred creatures go back millennia within Native American folklore and, in the global context, at least back to the *Epic of Gilgamesh*, and likely back to humanity's earliest moments. The modern field of anomalous primate studies and its parent discipline of cryptozoology began to coalesce in the late 1940s and early 1950s, when a rash of sightings and encounter stories appeared in the United States Pacific

Northwest and Western Canadian press, prompting a loose affiliation of journalists, outdoorsmen, and self-taught, amateur naturalists to investigate. This burst of activity found its stimulus in the sensation then being generated over reports coming in from Nepal about the Yeti, or Abominable Snowman. These monster hunters looked for scientific respectability and recognition for their work, despite being wary of the theoretical and advanced techniques of modern laboratory biology employed by the academic scientists also drawn to investigate the creatures. In a relationship sometimes close, and sometimes uncomfortable, the amateurs called for scientists to pay attention to what they were doing, and scolded them for ignoring their work. When academics did respond, the amateurs sometimes resented such incursions into their knowledge domains—especially when the "experts" contradicted their findings. Influential monster hunter René Dahinden often referred to scientists as "deadheads."[1]

The traditional heroic narrative of monster hunting situates mainstream scientists (the eggheads) as the villains rejecting the existence of anomalous primates and cryptozoology as something unworthy of study. The narrative gives a privileged place to untrained, but passionate amateur naturalists (the crackpots) who soldier on against great odds, including the unwarranted obstinacy of the mainstream against bringing knowledge of these creatures to light. The skepticism of some in mainstream science prompted pioneering cryptozoologist Ivan Sanderson to complain that scientists and other academics dismissed what he called ABSMs (Abominable Snowmen) without even looking at the abundant evidence. His *Abominable Snowmen: Legend Come to Life* (1961), held as a groundbreaking work in the canon, framed the critique of mainstream scientists' reactions to the evidence as much as it analyzed the evidence itself.[2] Barely able to control his pique, Sanderson quipped that "most of the skeptics are actually crackpots, yakking away in a vacuum of make-believe."[3] In his review of Sanderson's book, Johns Hopkins University anatomist William Straus said the same things about monster hunters as the monster hunters said about skeptics. Strauss framed the position most often embraced by members of the academic community. "The author's concept of what constitutes evidence," Straus said, "will scarcely be accepted by most scientists." He threw Sanderson's accusations back at him, arguing that Sanderson and his cohorts bore the burden of presenting acceptable proof. The evidence Sanderson put forward so enthusiastically, Straus said, "is anything but convincing."[4] Straus suggested that scientists would embrace acceptable evidence, if the amateurs possessed any.

Most commentators on cryptozoology found it easier to fall back on glib and superficially based insults rather than on thoughtful analysis. In his *Abominable Snowman* (1969), popular author Eric Norman rolled out the stereotype of the out-of-touch egghead. He said scientists needed to begin "emerging from their laboratories and moving out from behind their textbooks to take a sincere and penetrating look" at manlike monsters.[5] The *New York Times* focused on the stereotype of the crackpot, calling enthusiasts "little more than weekend amateurs, riflemen vowing to bring one back dead, an assortment of eccentric or unreliable individuals and obvious perpetrators of fraud."[6] With statements like these, the simplistic model for the relationship between amateur and academic over the study of manlike monsters dominated the field. The amateurs had an almost built-in wariness of the theoretical and laboratory methods of modern biology, zoology, and environmental science (a feeling less held now than in the past). They preferred the straightforward, hands-on approach of their field-naturalist forebears rather than theoretical thinking of scientific experts. To many amateurs the professionals lived lives of chair-bound academics doing their work in stuffy laboratories and libraries, knowing little of nature and not thinking the crackpot's real-world experience worth considering. Eggheads viewed the amateurs as unlettered devotees running around the woods, with no zoological training and collecting useless data, if they collected data at all.

Like all grand narratives, this one obscures a much more complex experience. Numerous academically trained scientists in the United States, United Kingdom, India, and Russia not only seriously believed anomalous primates existed, they actively pursued them, examined their physical traces, and worked out theoretical and evolutionary explanations for their existence. These academics encouraged the serious amateurs (those who endeavored to work in as rational and sober a way as possible), utilized their investigative impulses to supply materials for study, and listened to their points of view. The ranks of the amateurs contained a number of individuals who were intelligent, thoughtful researchers as annoyed as any university scientist by the eccentric trappings of some monster hunters. While the eccentricities of amateurs and the conflict between amateur and professional model dominate the discourse, a record also exists of cooperation between them.

The amateur naturalist monster hunters of the twentieth century find their antecedents in the first natural history enthusiasts of the Renaissance and Scientific Revolution periods. The history of the formation of mainstream nature-study disciplines began with amateur and later involved academic botanists, geologists, and ornithologists. Disagreement between the sides focused on issues of social status rather than scientific content. The nature of the physical material they studied was generally not called into question. With the later monster hunters the discussion came down to a conflict, not between two sides, but among three: amateur monster hunters who, while disagreeing on many aspects of their work, agreed the creatures existed in some form; academic scientists who agreed with them; and scientists skeptical of the entire enterprise, who disagreed not only with the amateurs, but with their believing colleagues. The monster hunters also faced the problem of the validity of the evidence they worked with. That no genuine "professional" or "academic" monster hunters resided among the ranks of professional scientists adds an additional analytic layer to negotiate in order to demarcate the factions. Those scientists who did engage the search for anomalous primates did not do so as part of their official duties.

If the modern definition of *professional* includes earning a living from the work one does, or having an affiliation with a museum or university, some of the amateurs rightly earned professional status as monster hunters over their academic counterparts. In their attempt to address the question of anomalous primate monsters, scientists suffered for their interests. Many academics who wanted to research anomalous primates did so quietly and tried to stay out of the public eye. They feared notoriety for their monster work would ruin their reputations. Despite this, archaeologist George Agogino did research and associated intimately with a number of the most ardent amateurs. Harvard anthropologist Carleton Coon followed the field closely and grew angry when dropped at the last moment from expeditions to Nepal to search for the Yeti. British primatologist William Charles Osman-Hill studied reports of unusual primates in Asia and supplied technical expertise to Yeti and Sasquatch hunts. His compatriot, John Napier, went out on a limb with his book, *Bigfoot: the Yeti and Sasquatch in Myth and Reality* (1973). Several Russian academics, including Boris Porshnev, pursued Asian anomalous primate reports, even receiving some state approval and funding. Others in the United States, England, India, and Russia had no doubt these creatures existed in theory and wanted to examine actual evidence, eagerly waiting for a breakthrough. By the early twenty-first century a growing coterie of university scholars, field biologists, government wildlife agents, and laboratory scientists around the world joined the fray. Of the band of academics who tackled the question of anomalous primates early on, Grover Krantz spoke the loudest and risked the most, operating as a public and energetic monster hunter. This book is unconcerned with proving whether Bigfoot is real or not. I leave that burden to others. I am concerned with what motivates scientists to look for such creatures. To do that this book is arranged in rough chronological order, but there are instances where I backtrack in order to cover certain topics more fully. Chapter 1 covers the amateur naturalist antecedents of modern monster hunters as well as of the founding thinkers of cryptozoology, along with an introduction to the life of Grover Krantz. Chapter 2 examines the case of the Abominable Snowman and the cold war connections of those scientists on both sides of the Atlantic who enthusiastically engaged in the hunt. How interest in the Yeti contributed to the growth of interest in anomalous primates in general, and in Sasquatch in particular, is examined. Chapter 3 introduces the main amateur naturalist monster hunters, along with the strange case of the Minnesota Iceman and its reception by the scientific mainstream. Along with the behavior of scientists in chapter 2, the reaction of scientists to the Iceman is instructive for an overall assessment of monsters and the mainstream. Chapter 4 is a detailed examination of the life and work of Grover Krantz, and how he developed an evolutionary explanation for the creature's existence. Chapter 5 lays out how mainstream anthropology reacted to the Paterson film. Chapter 6 examines the important aspect of evidence for manlike monsters, using the dermal ridge controversy and evidence from Russian monster hunters as case studies. Finally, chapter 7 rounds out the career of Grover Krantz and addresses the consequences for academics who take an interest in ideas from the fringes of scientific respectability. As with any historical analysis this book should not be taken as a final and definitive account, or as the only possible interpretation. I do not pretend to have covered every detail or individual involved.

The story of monster hunting and cryptozoology represents a cautionary tale for scientists and historians. It forces us to reconsider the role of the scientist and the subjects they investigate. The pursuit of monsters highlights the inherent difficulties scientists and historians encounter when they engage with anything from "outside the conventions and humdrum routine," difficulties that raise questions about which subjects are legitimate fields of study and which are not.[7] Examining the complex community of academic scientist and amateur naturalist anomalous primate researchers gives insight into the nature and function of scientific legitimacy, and into why some people chase monsters.

# Chapter 1

# Crackpots and Eggheads

*A monster is a thing deformed against kind both of man or of beast.*
*Mandeville's Travels* (1356)[1]

On a cold November in 1969 a gentle snow fell on the town of Bossburg, Washington, enveloping the area in forest quiet. Rumors had been rolling around the local populace during the Thanksgiving holiday about strange goings-on involving big, hairy shapes in the woods. Following up on the talk, local wilderness guide, trapper, and cougar breeder Ivan Marx came to Bossburg one evening to investigate and found Sasquatch tracks near the municipal dump. He had been alerted by butcher Joe Rhodes to keep an eye out for the tracks. Hikers, construction workers, trappers, and hunters had been finding such tracks all over the western United States and Canada, causing a stir in the local media. Rhodes also told Marx about a woman who stumbled across a creature lurking in the area the previous spring. Some sheriff's deputies arrived to look into the woman's story. They nosed about unenthusiastically accomplishing little more than exchanging snide remarks.

These new Bossburg reports, however, brought a small group of amateur naturalists, specialists in the study of hairy manlike monsters, descending upon the sleepy forest neighborhood of Bossburg and nearby Colville. With tracks at the dump, Marx called Canadian journalist John Green to come have a look. Green, a veteran Sasquatch researcher, was intrigued, but too busy with other business to come immediately so he called an acquaintance. This was Swiss-Canadian René Dahinden, who over the past decade had established himself, like Green, as one of the premier Sasquatch researchers. The Swiss-Canadian had a reputation as a no nonsense investigator grown wary of the techniques and motivations of many of his colleagues as well as of the dismissive behavior of academic scientists. He worked slowly and deliberately, having long since stopped running off at every mention of a sighting. He took a notoriously skeptical attitude—especially for a believer—toward any evidence that did not seem right. After a few days of consideration the Swiss-Canadian headed south across the U.S. border. Arriving at Bossburg, Dahinden joined up with Ivan Marx and others. Bundled up against the cold they fanned out over the area, crunching their muffled way through the ice and snow like the townsfolk from *Frankenstein* in pursuit of a monster. In the course of their searching they found over a thousand separate giant footprints. The tracks led off down a road, through trees, along a stream and out onto a field. At one point the tracks seemed to just step over a tall fence, then back again. The creature, obviously large and weighty, left tracks that broke through the brittle crust, sinking several inches. Boughs of trees covered in ice and snow had been easily pushed aside. The prints themselves looked vaguely human, yet different at the same time. In addition to their size, roughly 17 inches in length, the prints were exceptional in that the right foot had a strange pair of bulges on the outside edge and the middle toes seemed deformed. This unusual morphology prompted the searchers to refer to the unseen track maker as Cripplefoot. The Swiss-Canadian naturalist found himself puzzled by the prints, but the puzzlement created a growing sense that these might be the real thing. The tracks did not conform to any he had ever seen in his long experience. René Dahinden's normal reserve began to break down and his excitement grew at the thought of finally having found a large cache of genuine tracks and that the track's maker might still be hobbling about the vicinity.

As Sasquatch fever spread, more people came to Bossburg. Adventurers, schemers, and hucksters arrived and put together expeditions, made claims about a frozen carcass, film footage, a captured Sasquatch, and a severed foot. All of this undermined the investigation. The Swiss-Canadian became increasingly agitated and annoyed with the circus atmosphere. Eventually, a scientist arrived. A new member of the department of anthropology at Washington State University, he met with some of the researchers and saw a number of the remaining tracks before bad weather and gawkers obliterated them. He saw and inspected the fence the creature had stepped over. Eventually, back at WSU the academic received a pair of plaster casts of alleged Sasquatch handprints and then a good pair of footprint casts to study in detail. His knowledge of primate and human foot

anatomy convinced him the deformities of
dence it had been broken at one time and
prints and footprints had anatomical detai
ape. These details, he thought, could not
ers. Like René Dahinden, his amateur
Grover Krantz, started to think this migl
     Grover Krantz took the most promi
search for mystery apes of any physical an......,
both scholarly articles and popular books on the subject. He appeared
in news reports and on popular television shows, such as "Arthur C.
Clarke's Mysterious World" and "In Search Of," and helped found
the International Society for Cryptozoology. In his research into
the anatomy and physiology of these creatures, Krantz tried to do a
number of things, including proving their existence and giving scientific legitimacy to their study. Underneath all this floated a subtext
of wanting to take the study of manlike monsters out of the hands
of amateurs like the Swiss-Canadian and place it firmly in the hands
of anthropologists like himself, thus legitimizing it as a valid field
of intellectual inquiry. He wanted to stay away from anything even
remotely fantastic, preferring his work to fall within the bounds of
science and fearing being associated with the part of the monster
hunter community he considered the lunatic fringe.

Krantz worked on a number of issues other than the Sasquatch
during his 30-plus year career. His interests ranged from hominid
evolutionary mechanics and human population dynamics, to the
spread of language and culture, and the peopling of the Americas. He
consulted on the infamous Kennewick Man case and grappled with
creationists. Whether he wanted it to or not, however, the Sasquatch
work overshadowed all his scholarly efforts. It stood in reality as the
thing closest to his heart. He made all his work, whether consciously
or not, an effort to build intellectual, theoretical, and physical support for his theory that the Sasquatch descended from the Asian fossil
primate *Gigantopithecus*. He put his entire career into it and paid
a price as this obsession was not well received. He complained that
faculty members at Washington State University considered his work
an embarrassment and resisted his tenure and promotion. Early in
his career he told a magazine editor that the reactions to an article
containing some of his comments "have already damaged my professional reputation and exposed me to public ridicule by my peers."[3]
The men's adventure magazine *True* ran a profile of the life and work
of Peter Byrne whose book *The Search for Bigfoot: Monster, Myth or
Man?* had just been released.[4] The Irish-born Byrne was a big game

...rned Sasquatch and Yeti hunter who had been on numerous
...ions to find the creatures in Asia and North America. Krantz
...plained that the author, Al Stump, "filled in specific examples
...ich made me look like a fool."[5] Byrne commiserated with Krantz,
saying that much of the article and his quotes had been fabricated
by Stump. Being threatened by Krantz with a lawsuit for slander the
*True* publishers claimed Stump had quoted Krantz accurately and
that there was no libel or slander.[6] (*True* had a reputation for such
behavior; monster hunter Ivan Sanderson also claimed *True* editors
had mangled his statements and added parts that made it more sensational than he had intended). Not only did the article make Krantz
look less than scholarly to his scientific colleagues, it upset some of
the researchers studying anomalous primates. John Green, already a
legend in the Sasquatch community, wrote to Krantz to ask whether
he had really said the things the article said he did. The article had
Krantz praising the work of Byrne, whom Green did not care for. He
informed Krantz that, if he did hold those opinions, then "you may
consider your association with me at an end."[7] René Dahinden, who
also had a dark spot in his heart for the Irishman, told Krantz that
Byrne regularly made things up and "it is a question of dealing with
known liars and conman [sic] and make believe artists."[8] The article
had long-term effect on the community and generated animosities
that lasted into the twenty-first century. This experience joined many
similar ones in Krantz's career. He often found himself on the receiving end of criticism from friends and enemies on both the amateur
and academic side of the issue. Later, Krantz claimed that a career
chasing Sasquatch had exacted a price from him in "time, money, and
professional reputation, not to mention social ridicule."[9] A good portion of the mainstream world of science either dismissed, or simply
ignored, his work.[10]

René Dahinden, the Swiss-Canadian at Bossburg, also became
Krantz's nemesis. An autodidact and self-made monster researcher,
Dahinden had clawed his way up from poverty to something slightly
above poverty to make the search for the Sasquatch the central con-
suming force of his life. He resented "experts" with their formal training and theoretical ways of thinking and soon developed a dislike for
Grover Krantz. During one of their many disagreements, he sputtered
at Krantz, "every time you open your mouth to the press you make a
bunch of downright stupid statements." Dahinden then threatened,
"I will pull you down and blackball you in the Sasquatch research."[11]
In later years another Sasquatch enthusiast growled contemptuously
that "more than anything Krantz's entrance into Bigfootery [sic] was

a death blow" to the field, alleging Krantz had a tendency to fall prey to hoaxers, which in turn soured relations with mainstream science.[12] In a review of Krantz's *Bigfoot/Sasquatch Evidence* (1999), author Daniel Perez, while calling Krantz "the most widely recognized academic authority on the subject," found some of his work "questionable." He said Krantz had accepted as genuine a set of track casts that had been intentionally faked in order to test Krantz's mettle. "He took the bait," Perez lamented, "and endorsed them as real."[13]

The Cripplefoot tracks found the winter of 1969, and the impact they had on the field, are emblematic of the hunt for anomalous primates. The Bossburg incident stands as a microcosm of the place of strange and unusual evidence in the history of science. What is to be made of such a discovery? Should it be investigated and taken seriously or dismissed as obvious hokum? When the amateurs, René Dahinden and John Green, and the academic, Grover Krantz, stood discussing the Cripplefoot in the snow of Washington State they unconsciously engaged in a long running and crucial discourse, not over monsters, but about the relationship between amateur naturalists and professional, academic scientists; about the demarcation of science and pseudoscience; and about the location of the seat of intellectual authority: was it with amateurs or academics, crackpots or eggheads?

## What Is a Scientist?

Though few of them realized it, the monster hunters swarming Bossburg could trace their lineage back to the eighteenth century. A long history of amateur naturalist activity preceded the events in the Pacific Northwest. For most of Western history the idea of a conflict between professional and amateur science would have struck participants more than two centuries ago as incongruous, since the role of professional did not exist. Even the most serious, scholarly, and dedicated investigators studying the natural world did so as amateurs driven by a desire for intellectual attainment. They studied minerals, animals, the movements of the heavens, birds, other human beings—a whole host of subjects, although they rarely had formal scientific training. Except for occasionally selling objects they gathered to collectors, they received no monetary remuneration for their work and, instead, pursued science and nature out of a deep inner passion, an avocation rather than a vocation. Far from seeking professional status, a resistance to professionalism permeated the work of early naturalists. Well-to-do British science enthusiasts for example tended to shun the idea of being labeled professional, as that implied they made a living as tradesmen; something repugnant to gentlemen.[14]

The existence of a class of learned individuals studying what today is called science goes back to antiquity, as does fascination with the more unusual aspects of the natural world. Aristotle wondered about reports of exotic creatures. Pliny and other classical authors tried to distinguish reality from fantasy when it came to legends of zoology. By the Middle Ages, and then the Renaissance, learned Europeans pondered animals from far-off lands and became fixated on "monsters." The term could incorporate exotic and fanciful animals and plants as well as people. Indeed, a good bit of medieval natural history writing concerned such organisms. Illuminated bestiaries based descriptions of animals and people less on careful examinations of the animals themselves than on myths, legends, and a faulty knowledge of non-European life forms. Ironically, Aristotle, who argued the investigator should place emphasis on direct observation rather than on past authorities, became the preeminent authority relied upon by European scholastics. This preoccupation with classical authors over empiricism in many ways retarded medieval learning. It tended to push most early naturalists into taking old and fantastic stories at face value rather than investigating them. By the early sixteenth century, however, preoccupation with fantastic creatures lessened, and the character of the out-of-touch scholar was increasingly replaced by that of the field naturalist. Reliance upon myth, legend, and past authority was challenged by direct observation and study. It began with the science of describing while engaging in what practitioners considered an improving pastime rather than a profession, and certainly not undertaken out of any monetary concerns. Enthusiasts' cabinets of curiosity bulged with rocks, minerals, fossils, and taxidermy wonders as well as animal skeletons, while their greenhouses brimmed with exotic plants. Some collected these objects and showed them off for their intrinsic physical beauty or strangeness. Others went beyond that and carefully cataloged and arranged them with growing sophistication. For others still, these activities included ponderings on larger zoological and biological questions and became more theoretical.[15]

Across the ocean, well into the mid-eighteenth century, no colleges in North America taught science in a comprehensive way, if at all. A medical degree came closest to science training, and that often had to be acquired abroad. This did not stop interested persons from fanning out over the countryside to marvel at and collect examples of the natural world or stare up at the heavens. Explorers included both American and European naturalists eager to find new and wondrous things. The ranks of these early naturalists were filled out with doctors, lawyers, and landed men, all with a passion for nature and running the gamut from dilettante to serious investigator. It was easy to find subjects to study. So many new species of plants and animals presented themselves that one could hardly walk down a country lane without finding a new flower or plant that had yet to be recorded. Itinerant botanists could make a reputation by supplying seeds to European collectors who developed a mania for exotic American plants and flowers to grow in their gardens. Besides the species already known to Europeans, material the Old World had never seen grew copiously. Many claims of exotic American creatures and growing things thrilled and perplexed Europeans. Some species—like the moose—seemed so outlandish they were considered myths. (The native peoples, of course, had long known of these creatures and even collected fossils for ritual objects.)[16]

## American Science Transforming

As in Europe and England, distinct scientific disciplines emerged in America as some naturalists focused on botanical collecting, others on zoological or ornithological interests, and some on the earth sciences. Some drifted toward ethnology and the question of the nature and origins of the aboriginal peoples.[17] Nineteenth-century anthropologists like E. G. Squier, Josiah Priest, and William Pidgeon undertook extensive surveys of the innumerable artificial mounds that dotted the landscape from New York to Ohio. The unusual, exotic, and mysterious mounds baffled naturalists as to their origins. To explain these unusual structures, amateur ethnologists and archaeologists developed theories from the reasonable to the outlandish, all the while developing the rudiments of excavation techniques.

The most popular theory of the origins of these structures, the myth of the Mound Builders, postulated a mysterious people, separate from and superior to the Indians who were all but exterminated in a great conflict with those same "degenerate" aborigines. This idea dominated anthropological thinking in America for decades. Ultimately, the myth of the Mound Builders gave way to the work of a growing cadre of professional and semi-professional scientists and academics engaged in a general professionalization of American science, which in turn became associated with the new government sponsored Smithsonian Institution, founded in 1846, and the Bureau of American Ethnology and the U.S. Geological Survey, both of which began operations in 1879. Since the American Revolution the

federal government as well as state governments had increasingly supported scientific expeditions and research. The opening of these government agencies meant more money for science. Initially, amateur
naturalists took full advantage of the largesse, though the availabil-
ity of such funds could be sporadic, and some had misgivings about
the government being involved in scientific endeavors.[18] Increasingly,
the growing ranks of academics took advantage as well, to the point
where it became difficult, though not impossible, for amateurs to
receive federal funding. By the middle of the nineteenth century
the rift between traditional amateur naturalists and growing professionalization began to show and was pushed along by institutions of
higher learning. Schools such as Yale and Harvard, later Princeton
and then Johns Hopkins, introduced science curricula and research
programs. Nature enthusiasts no longer need be self-taught. At Yale,
the Peabody Museum, and at Harvard, the Museum of Comparative
Zoology, promoted serious research along with the more traditional
museum functions of collection and exhibition. The Smithsonian
established the peer-review system to vet articles and research for scientific efficacy and to promote community standards for scholarship.
All this helped institute scientific disciplines with shared ideals and
goals, so ideas could be discussed and information shared within the
communities of scientific practitioners.[19]

Along with the aboriginal peoples, birds and fossils were areas of
great interest in nineteenth-century America. Like botany, ornithology began as a pastime that followed a culture of collecting. In the
1880s academic ornithologists began to exert influence over bird
watching by enforcing species-naming procedures and other forms of
standardization.[20] Fossil hunting too began as a pastime. By the later
nineteenth century academic paleontologists came to dominate the
field as far as the interpretation of materials. The increasing reliance
upon evolution theory to explain the fossil record—and the atten-
dant growing conflict between theology and secularism—contributed to the continuing professionalization of the field. The number
of paid positions available for naturalists in museums, universities,
and government also grew. This helped shift emphasis on natural history and earth studies away from the self-sustaining virtuosi. The
amateur fossil hunters (both secular and religious) did not readily
relinquish their standing. One place the amateur fossil hunters still
held sway was in the hands-on collecting of fossils, particularly in
the desolate and sometimes dangerous wilds of the western territo-
ries. Museums and universities relied on collectors to supply them
with specimens for study. The fossil "Bone Wars" and "Bone Rushes"

conducted by scientists in their laboratories and collectors in the field resulted from this relationship. The amateurs—ranchers, cowboys and mineral prospectors—engaged with professors, not only over control of the fossil sites themselves but also over interpretation of the fossils. As scientists depended on the collectors, they alternately argued with and placated the amateurs. Despite their lack of formal training, several amateur collectors rose to respected status among the academics.[21] One of the most powerful and well known scientists in America in the late nineteenth and early twentieth century, Henry Fairfield Osborn (1860–1935), president of the American Museum of Natural History in New York (a primary market for fossils from the West), encouraged a number of the amateurs who supplied his fossil needs to go on to obtain professional training.[22] So, while academics did manage to create the perception that they were in charge of the biological and earth sciences, they did not push their rivals out completely (they neither wanted to, nor would the amateurs let them). They utilized the skills, experiences, and enthusiasm of the amateurs to build up enormous compendiums of biological data. While the new class of academic scientists did not reject field work, they needed to do extensive laboratory and library work as well. This contributed to the growing public perception of the difference and demarcation between amateur naturalists and professional scientists: the amateurs worked in the field, while the professionals worked in the laboratory. The academics took the previously ad hoc and haphazard collecting impulse of the amateurs and attempted to organize it under their leadership.[23] The professionals sought to control rather than eliminate the amateurs.

Over the course of the nineteenth century, leadership in the study of the Indians, birds, rocks, and fossils, as well as medical practice shifted from the hands of self-financed, self-motivated amateur naturalists to the hands of the work-a-day, government, museum, and university-affiliated academic scientists. A pattern asserted itself whereby knowledge domains begun and dominated by amateurs came under the control of professionals and academics. Lone virtuosi and amateur natural history associations continued to exert influence locally and helped facilitate transfers of knowledge, but a dominant academic archaeological and zoological nexus formed in places like the Smithsonian Institution in Washington, D.C., at the American Museum of Natural History and Columbia University in New York, at Harvard in Boston, Yale in New Haven, Johns Hopkins in Baltimore, the University of Chicago, and later at the University of California, Berkeley, where Grover Krantz studied as an undergraduate.

The change in the underlying structure of American science can be seen not as an abrupt revolution in which professionals ousted amateurs completely, but a subtler and nuanced evolution. If a biological cladogram were drawn showing the relationship between amateur and professional, it would be the professionals who branched off from the strong amateur line, while the amateurs continued on, though in a diminished capacity as leaders. Amateur status, held up as a mark of an erudite intellectual and man of society in the seventeenth and eighteenth centuries, increasingly was seen as a liability to professional advancement in the nineteenth and twentieth. Those not associated with a museum, university, or the federal government, especially in natural history studies, increasingly found themselves on the outside. At the same time, while the academics began to overshadow the amateurs, the professionals found themselves more and more an elite outside of society viewed alternately with awe and suspicion because of their lofty position and authority.[24]

Organizing and defining these workers has proven no easy task for historians. Nathan Reingold's study of the origins of modern science resorted to the term "cultivator" in order to get around the thorny professional scientist/amateur naturalist model. Reingold saw North American scientists of the nineteenth century as falling into three categories: vocational, avocational, and research scientists.[25] Adrian Desmond points out "even getting an embedded, localized definition of 'professional'…is itself becoming *the* problematic in the history of biology."[26] Historian Paul Lucier argued that the term "professional scientist" was problematic in the nineteenth century context. Professionals "were men of science who engaged in commercial relations with private enterprises and took fees for their services." Scientists held positions with institutes of higher learning and worked for pure science unconnected to commerce. There have been many and contradictory arguments over the meaning of what constitutes amateur, professional, scientist, and as Lucier puts it, the most contradictory of them all "professional scientist."[27]

The monster hunters, too, broke down into groups based upon doing their work purely for the love of it or as a living, yet resist categorization. Those making a living from their monster work being the smaller population. Within these there are important subgroups: amateur monster hunters who believe the creatures exist in some form, academic scientists who agree with them, and scientists skeptical of the entire enterprise who disagree not only with the amateurs, but their believing colleagues. The amateur camp can be subdivided yet again into those who think Sasquatch is a genuine biological animal, those who believe it is a spirit entity, and those who believe it to be an extraterrestrial. These sub-groups cross the boundaries of the vocational and avocational amateurs. Complicating the situation more, a number of the monster hunters are difficult to even label as amateurs. These individuals have no research focus—even for the world of monster hunting—and are more adventurers and sportsman. They enjoy the hunt rather than the learning. More akin to bird watchers than ornithologists, their goal is to see the creature and prove they did. Beyond that they have little interest in the research aspects of what they do. They can be termed "monster watchers." Yet another category could be "elite amateur" monster hunters. These individuals not only make a living from their work, but do so in the more intellectual realm of writer-theorizer rather than as field guide or tracker. These could be seen as professional cryptozoologists. That the academic world has no position for cryptozoologist at the moment makes this last statement problematic. It also requires the definition of another term.

## What Is a Monster?

While the serious researchers at Bossburg came looking for an animal, others came looking for a monster. The English word comes from the Latin *monstrum*, meaning portent or prodigy. Several characters in Greek and Roman mythology received the label, not because of their physical appearance, but because their arrival presaged things to come. The Greek synonym is *tera* and is the root of the word teratology: the study of monstrous births now known as birth defects. The appearance of such deformities in Europe during the Middle Ages and Renaissance frightened people into believing them to be omens of impending doom. Because such birth defects went beyond the expected of a newborn, the term monster—also "wonder"—took on the added meaning of something outside the normal, something ugly and fearsome; something to be avoided, yet fascinating and worth studying by learned persons.[28]

The interpretations and perceived meanings of such monsters varied in complexity. They could be taken as signs of the perfidy and ungodliness of an individual—usually blame fell upon the mother—or that of an entire town or nation. They could alternate between signs of God's power, mercy or displeasure. In Western Europe a baby with multiple arms and legs often generated fear among local people. In India a baby born with multiple arms and legs could be seen as resembling the God Shiva and so might be seen as a good omen. The notion of a monster as a terrifying beast also began to form during the Renaissance when the darker connotations of monstrous births were attributed to entities that inhabited the unknown and forbidding parts of the world as well as of the human psyche, particularly the wild men and what came to be known as the monstrous races. Legends about strange creatures and bigoted—or simply misunderstanding—descriptions of foreigners led to a long list of monstrous races. The Blemmyea (men with their faces in their chests), Sciopods (men with one giant foot), Cynocephali (dog headed men), and others were popularized by first-century Roman encyclopaedist, Pliny, then copied and recopied in manuscripts for a millennium after.[29] These monsters also went through changes of their own.[30] Some, like the Blemmeyea, went extinct. The Cynocephali, too, dropped out of sight while their cousins the werewolves hung on precariously. Various sea monsters came and went. Nineteenth-century naturalists sometimes still used the word monster, but to mean a hereditary mutation. The growing reliance upon scientific authority put a strain on belief in monsters. By the twentieth century monstrous creatures had virtually disappeared in the industrialized world outside of legend and fiction. Just when all such monsters seemed to have been vanquished, they found a second life for a new generation. By the twentieth century two-headed babies no longer generated the awe and dread they once did. The science of teratology had shown the genetic and pathological causes of such unfortunates so they became more crudely freaks used for entertainment rather than for moral wisdom. Sciopods, mermaids, and werewolves seemed quaint and old fashioned. The manlike monster and anomalous primate now stepped onto center stage as preeminent exemplars of strangely formed human bodies, and along with them a new field of study arose.

## Cryptozoology

By the middle of the twentieth century, the difference between professional, academic scientist, and amateur naturalist seemed no clearer. Even though they had been supplanted as leaders in the scientific world, their amateur collecting and observational impulses continued unabated. A considerable population of enthusiasts, as serious as ever about their work, cultivated gardens and aquaria, built their own telescopes and observatories, and published in their own journals and magazines.[31] Popular publishing helped fill the need for a place for amateurs as well as the interested general public to discuss science. Magazines like *Popular Science, Popular Mechanics, Practical*

*Mechanics, Galaxy* and others catered to readers with a wide variety of interests. An almost unquenchable thirst for scientific knowledge led to a boom in publishing geared to this market in both North America and the United Kingdom.[32] The majority of amateurs pursued their interests as a private enjoyment and passion. Some still wanted to, and did, contribute to mainstream scientific knowledge. Amateur astronomers, for example discovered and studied heavenly bodies and observed celestial phenomena through their often-sophisticated homemade telescopes. Bird watchers still watched, and rock hounds continued to discover new fossil species.

For serious amateur researchers who yearned to make genuine contributions to science, particularly in a realm of natural history all but abandoned by academics, manlike monsters seemed real animals which, while extremely rare, could be made well known simply by dogged persistence and empirical study. By looking into the question of anomalous primates these individuals insisted they were not chasing phantoms, and bristled at the accusations that they were involved in something paranormal. The argument over what constituted a "real" scientist continued with the monster hunters. The early proponents of the field—the ones who teetered precariously in the twilight between professional and amateur—stepped forward as the most vociferous in their assertions. Such amateurs found an outlet for their energies in the emergent field of *cryptozoology.*

The operative concept of cryptozoology (from the Greek for "hidden animals") is that some animals thought to have gone extinct still lurk in remote parts of the world, and that creatures once thought only to have existed as myths or legends have some basis in biological reality. For a field dominated by amateurs, cryptozoology began with men who considered themselves scientists. The founding intellectuals of the field were Scottish naturalist Ivan Sanderson and French zoologist Bernard Heuvelmans.

Ivan Sanderson (1911–73) already had a career as a writer of nature and zoology books geared to the popular market, including *Anthology of Animal Tales* (1946) and *The Monkey Kingdom* (1957), when he turned seriously to mysterious animals. Born in Edinburgh, Sanderson attended Eton and Cambridge. Entering Eton in 1924, Sanderson was tutored by C. J. Rowlatt, graduating in 1928.[33] Sanderson traveled extensively with both his mother and father and acquired a taste for travel and the natural world. He entered Trinity College, Cambridge, in 1929 at the age of 18, studying zoology and botany under the Natural Sciences Tripos (an honors BA), graduating in 1932.[34] Following graduation he worked

as a demonstrator at Cambridge until his mentor, Professor Clive
Forster-Cooper (1880–1947), suggested he find his destiny in field
research. A paleontologist, Forster-Cooper (later Sir Clive) became
greatly enamored of life in the field. He left Cambridge to become
director of the British Museum, Natural History, in 1938 and held
that post until his death. Forster-Cooper had discovered field work
at a young age traveling to the Maldive Islands, North Africa, and
then to North America to study fossil mammals under paleontologist Henry Fairfield Osborn, soon to be president of the American
Museum of Natural History.[35] Forster-Cooper passed on his love of
the field to Sanderson, who took his advice and became entranced
by the excitement and romance of field work, never returning to
formal studies. Despite having begun work toward a master's degree,
he did not receive it until years later. This lack of postgraduate training haunted him and motivated a defensiveness about his status as a
scientist that he never quite shook off.

Sanderson traveled extensively, including leading an expedition to
Africa for the Linnaean Society in 1932, and later pioneered television nature programming. His life took on the appearance of one long
zoological expedition, even while he acted as an intelligence agent in
the Caribbean for the Allies during World War II. He eventually took
up permanent residence in the United States, buying a farm in New
Jersey where he could indulge his love of animals.[36] This attraction
could sometimes be a problem, as in 1951, when living in New York
City, his pet monkey got loose and bit a neighbor. Sanderson's wife,
Alma, had to appear before the Department of Health to explain and
have the animal inspected.[37]

Making his living as a full-time writer, Sanderson increasingly
focused on subjects at the fringes of science, including Unidentified
Flying Objects (UFOs). He ran a series of research organizations,
and employed assistants. He first drew attention for his work on
anomalous animals through a series of articles in the late 1940s and
early 1950s, which appeared in various family and men's adventure-
oriented magazines. One of these articles arguably helped launch the
field of cryptozoology.

Born in France, but raised in Belgium, Bernard Heuvelmans (1916–
2001) earned a doctorate in zoology in 1939 from the Université
Libre de Bruxelles, his doctoral thesis concerning Aardvark teeth,
although the mythological aspects of zoological history fascinated
him more. Heuvelmans began to think some fantastic animals had
a basis in fact. He spent years quietly collecting reports and other
materials about such animals. Unable to find work as a zoologist in

war-torn Europe, Heuvelmans worked variously as a jazz musician and comic performer as well as a writer. In 1948 he read the *Saturday Evening Post* article by Ivan Sanderson, "There Could Be Dinosaurs," in which the author suggested local legends supported the idea that small populations of extinct animals had survived to modern times.[38] Seeing someone else was thinking along the same lines, Heuvelmans began to consider publishing his own work and going back to the career of his training. The result of this early period of wider research and inspiration was *Sur la Piste des Bêtes Ignorées* (1955; released in English in 1958 as *On the Track of Unknown Animals*). The book established Heuvelmans as an authority on the more unusual aspects of zoology and became the standard text to judge other researchers by. Though Heuvelmans's book covered a wide range of unusual animals, he argued—albeit tentatively—for a connection between the Nepalese Yeti and the extinct giant Asian fossil ape *Gigantopithecus*. More broadly, *On the Track of Unknown Animals* took its place as the first widely influential work in the field of cryptozoology, inspiring legions of followers and intellectual children.

Heuvelmans authored many books on unusual animals, including the seminal *In the Wake of the Sea Serpents* (1968), in which he gave credit to the late nineteenth-century Dutch zoologist, Antoon Cornelis Oudemans (1858–1943), as the unsung father of the field of anomalous animal research. Though there had been many articles and pamphlets in England, Europe, and America on sea monsters, Oudemans's *The Great Sea Serpent* (1892) was the first modern work to take such creatures seriously and to approach the topic in a scholarly fashion. As director of the Royal Zoological and Botanical Gardens at The Hague, Oudemans became fascinated and perplexed by the many sea monster stories then prevalent. He collected over a hundred published accounts and laboriously analyzed them. His methodology of eliminating obvious misidentifications and hoaxes in order to find a core of reliable descriptions became a major part of Heuvelmans's approach. While Oudemans never gave a label to his work, later cryptozoologists see him as an originator of their field thanks to Heuvelmans's resurrection of his memory.

Though Heuvelmans argued the field of cryptozoology began in the nineteenth century, for the purposes of this study the starting point of this discipline will be taken as the mid-twentieth century and the introduction of the term cryptozoology. The first published appearance of the word may be in Lucien Blancou's *Géographie Cynégétique du Monde* (1959).[39] Heuvelmans originally attributed its use to Ivan Sanderson. In 1968 Heuvelmans wrote that "when he

[Sanderson] was still in school he invented the word 'cryptozoology'; or the science of hidden animals, which I was to coin much later, quite unaware that he had already done so." By 1984 Heuvelmans claimed, "I actually coined the term cryptozoology," without any mention of Sanderson's role.[40] In his work, Heuvelmans argues that the search for *cryptids* (a term first used in 1983 by Canadian John Wall) must be thorough, rigorous, and scientific, since the object is to look not only for physical animals in the field but also for the folkloric nature of such creatures.[41] Heuvelmans insisted the cryptozoologist must plow through the mountains of artwork and legends that wrapped the animals like cultural camouflage. He helped ingrain in the later monster hunters the importance of being as careful and scientific as possible. As a trained scientist (unlike most of his followers) Heuvelmans wanted his unknown animal work to meet a high scholarly standard. He considered this new endeavor as intensely interdisciplinary, employing the biological sciences, mythology, linguistics, history, and art history in a holistic approach.[42]

Just as important and consequential for the study of cryptids is the sense of intellectual conflict with the mainstream that Heuvelmans injected into the discussion. He argued that individual naturalists back to eighteenth-century Europe wrote about monsters, attempting to explain them in zoological, not supernatural, terms. Resistance to such work began, Heuvelmans claimed, in the nineteenth century when "armchair naturalists most of the time—eventually reacted with indignation, exaggerated alacrity and sometimes violence" against such ideas.[43] Deeply annoyed by what he perceived as mainstream science's less than enthusiastic response to cryptozoology, he knew where things had gone wrong. Heuvelmans believed the drift began with the late nineteenth century professionalization of a scientific establishment. Masquerading as Enlightenment naturalism, Heuvelmans claimed, professional science "borrowed from some religious creed." A formal hierarchy appeared, which like the medieval church, would brook no opposition or dissent. George Cuvier, in France; Richard Owen, in England; and Rudolf Virchow, in Germany, set themselves up like cardinals passing judgment and controlling thought. Only "a few enlightened laymen" dared suggest there might be something to myths and legends of fantastic creatures. It should not be seen as unusual that Heuvelmans would take this tack. Drafted at the outbreak of World War II he lived through the Nazi invasion and occupation of his French homeland and his adopted home of Belgium. Captured, he managed to escape and spent the rest of the occupation dodging Nazi police.[44] At the same time he worked as a writer for *Le Soir*, a Belgian newspaper that had been taken over by the occupation forces. While at *Le Soir*, Heuvelmans met and befriended the Belgian artist George Remi, who under the pseudonym Hergé, produced the wildly popular Tin Tin comic books. The two remained friends despite the accusations that Hergé had been a Nazi sympathizer, and Heuvelmans acted as a technical consultant on a number of Tin Tin stories.[45]

His hatred of the Nazis contributed to Heuvelmans's ingrained "us versus them" attitude toward the "dictators of science," which would have a long shelf life in anomalous primate studies. He established the trope of the free thinking, unfettered amateur monster hunter fighting against the dark forces of scientific mainstream conservatism, but he wanted to have his cake and eat it too. While lamenting the dismissive attitude of the mainstream in *On the Track of Unknown Animals* and *In the Wake of the Sea Serpents*, Heuvelmans listed the many mainstream journals that carried monster-related articles. He also referred to legendary scientists like Joseph Banks and T. H. Huxley as being sympathetic to monster studies. These men, however, struggled in the complex social interaction that was the professionalization of British science, helping build the fledgling establishment as leading arbiters of scientific thinking, just the kind of men Heuvelmans said kept interest in monsters down.[46]

To be fair to Heuvelmans, his critique hit close to the mark: he just did not seem to understand why. The ideas of Charles Darwin and his followers helped power the late nineteenth-century turn in natural history toward professionalization. Evolution science played a complex role in the discussion of monsters and monsters a complex role in early discourse on evolution. Some who objected to evolution pointed to mermaids and other similar composite creatures and chuckled that to believe in evolution one must believe in such childish and superstitious phantasms. Opponents argued mermaids, the Minotaur, and sea serpents imaginary so to prove evolution, scientists had to produce examples of such beasts. As these creatures did not exist evolution must be a sham. Detractors challenged that, if Darwin and his followers were correct, then the world should teem with such composite hybrids. Evolutionists countered that the archaeopteryx, no less fantastic than a hippogriff, had proof in the fossil record. The dinosaurs and long-necked plesiosaurs roamed the land and swam the seas no less majestically than anything from Greek mythology. Some evolutionists argued that if sea serpents did exist, evolution supported their origins. Most scientists, however, dismissed monsters as charming creatures of folklore and superstition that Darwin helped do away with. Evolutionary biologists felt their work brought a deeper and more profound understanding of the workings of nature so that people would be less likely to believe in the existence of mermaids and other wonders. [47] While monsters still generated popular interest, less and less scholarly attention was paid to them by naturalists, who held evolution as an overarching paradigm. Scientists rejected monsters as real because the growing professionalism of science and its reliance upon evolution theory showed that a mermaid not only did not exist, but could not exist. Any attempt by scientists to engage with the concept of monsters could therefore be seen as an intellectual step backward.

This heroic narrative of science and evolution banishing belief in monsters is not quite as simple as it might seem. While evolutionary biology certainly rejected the notion of the mermaid (half fish-half human) and the werewolf (half wolf/dog-half human), it had a far more complex opinion on the relationship of primates and humans. Indeed, the very notion of human evolution is built upon a close relationship to primates as the origins of *Homo sapiens*. In this way evolutionary biology seemed to give tacit support to ape-manlike monsters, whether the mainstream wanted it to or not. Regardless of whether anomalous primates existed, science grudgingly said they could at least in theory, and so the Yeti had a much less formidable hurdle to negotiate to be accepted as biologically genuine as almost any monster previously had. The problem the mainstream science community posed, particularly that of paleoanthropology, involved a willingness to accept the primate-related origins of humans, on one hand, but difficulty accepting that living ape-men might be lurking about, on the other. This seemingly contradictory attitude infuriated Heuvelmans and Sanderson.

Inspired in turn by Heuvelmans, Sanderson published a series of articles on wild men in the Italian zoology journal *Genus* in the early 1960s. Published in several languages, this journal's interests included genetics, population studies, and the disgraced concept of eugenics. Indeed, the journal served as the house organ of the Societa Italiana Di Genetica Ed Eugenica, whose founder, the fascist theorist and statistician Corrado Gini, also had an enthusiasm for the Yeti. In the first of his articles Sanderson surveyed medieval texts for references to wild men and monstrous races and argued that these depictions represented "relic knowledge of some fully haired primitives or sub-humans that once inhabited Western Eurasia." [48] He felt references to hairy humanoids in medieval folklore formed a kind of folk memory of proto-humans like *Pithecanthropus* and the Neanderthals, which survived into the modern period of *Homo sapiens*. "It is our contention," he said, "that the Wudewása are detailed and accurate descriptions of Neanderthals...that lingered on in Europe...until comparatively late dates."[49] The last pockets of these relics roamed the wastes of Russia and accounted for the sightings and legends of the Sasquatch-like *Almasti*.

Sanderson did not think all manlike monsters the same. His interpretation of the evidence showed a fairly wide diversity and dispersal among these creatures. Sasquatch, which could be traced back through centuries of Native American folklore and belief, constituted a species separate from the Russian or Nepalese variants. He agreed with author Harold Gladwin who, in *Men Out of Asia* (1947), suggested waves of hominids had passed into the Americas. One of these waves, Sanderson suggested, included the Sasquatch.[50] The real impact of Sanderson's theories hit the amateur Sasquatch community, not through these articles, but through the book-length version he produced at the same time.

In 1961 Sanderson published *Abominable Snowmen: Legend Come to Life*, which recapitulated the work he had done on the topic up to that point. Though the Yeti formed the core of the book, he also discussed related creatures. Sanderson admitted in his typically jaunty writing style that when he first heard reports of Yeti-like creatures in California in 1958 he thought the idea "sounded quite balmy."[51] Later, however, he came around to accepting the existence of such a creature. He believed several different Abominable Snowmen (ABSMs, a generic term for any anomalous humanlike creature) roamed the world. Some might be Neanderthal holdovers, some *Gigantopithecus* relatives, some *Pithecanthropus* and *Sinanthropus*, and even some feral humans. Sanderson noted that reports of wild men and ABSMs from China came from the same regions in which *Gigantopithecus* fossils had been found. This "would bear out Bernard Heuvelmans' theory that they are indeed *Gigantopithecus*."[52] He did quite a bit of research on the creatures, talking to reputed eyewitnesses and collecting historic texts.

Along with the work on the Yeti, Sanderson like Heuvelmans let loose with a tirade against scientists. While he considered himself an academic, others dogged Sanderson throughout his career with accusations of his being little more than a popularizer and amateur. His outward air of enthusiasm and nonchalance masked self-consciousness about his lack of scientific graduate credentials. As a result he attempted to separate out the "true scientists" from the "experts." The true scientist, himself naturally, should not be judged by fancy degrees but by a willingness to see new things in new ways, unafraid of bucking the status quo, not tied to any hierarchy. He did not think "experts" very intelligent, either. "The amount of plain 'ignorance' even among the most learned is quite terrifying," he said. Then, not wanting to seem too cranky, he continued that "though the truly learned are always the first to admit this." Going on, he said, "We have no quarrel with the learned nor with the true scientist: our clash is with the so-called 'experts.' "[53]

Sanderson considered the search for ABSMs as more than just the search for monsters. He saw larger implications in the battle between monster enthusiasts and mainstream academics. When "The People" call for something, like investigating manlike monsters, he said, scientists are supposed to respond to their will. If they do not, if they say such a thing is a waste of time or makes no sense, then an erosion of our democratic society has occurred. Scientists have a duty to investigate at The People's bidding and to appear in the court of public opinion to explain themselves. All mainstream scientists ever did, according to Sanderson, was rail and rant against the idea of cryptozoology and assert it was foolishness. "Thus when The People appealed to the Experts in their guise of 'Scientists' the court [of public opinion] was subjected to a tirade," by them even worse than that of the Nazis.[54]

With democracy hinging on proof of the existence of the Yeti, Sanderson also linked the search to the cold war. In an unpublished manuscript titled "The Race for Our Souls," Sanderson laid out what he thought the consequences would be should the Soviets find the Yeti first. Sanderson raised the hunt for Sasquatch and Yeti to global proportions. The Communists held the lead not just in science, as shown by their success with Sputnik, but by leading the way in ABSM research. They had, he said, devoted many scientists and an entire scientific building to the problem. "In fact," he intoned, "they seem to think of everything and they appear to be a lot more pragmatic and a lot less squeamish than we are." The subtext had the Russians receiving state funding while he and his compatriots barely scraped by in need of either government funding or rich benefactors to bankroll the work. The reason the Soviets put so much effort into this—and why the West needed to beat them at it—was that if the Communists found a living Yeti it could be used by propagandists to "rock the entire religious and ethical pyramid [of the West] to its very foundations." In his apocalyptic vision and shaky logic, knowledge of the existence of ABSMs in the wrong hands raised questions about human evolution *and* religious orthodoxy. Just why this would not happen if the West found it, Sanderson did not articulate. He argued, quite seriously, that the very future of freedom and democracy itself, possibly all life on Earth, rested on who found a Sasquatch first.[55]

*Abominable Snowmen* is, in fact, two books: one the exposition of the evidence Sanderson thought proved creatures like the Yeti existed, and the other a barely controlled rant against those who disagreed with him, especially academics and "experts." "Besides being dull, most professional skeptics are insufferably conceited" because they have not looked at the material.[56] Sanderson was sensitive about being labeled an amateur. He felt his education, past work, and publications more than made up for his not having a doctorate or academic affiliation. He argued contemptuously that the term scientist only referred to where someone worked, not to their skills, intelligence, or scholarly output. "Thus anyone who is not employed in or by very certain specific categories of organizations," he said, are referred to "often scathingly so as 'an amateur.'"[57] Harking back to the early period of natural history studies, he suggested that "real" scientists eschewed connection to a university or commercial interests. For Sanderson, as he did not have them, such connections undermined an individual's status as a scientist. It became a common tactic for amateur monster enthusiasts to say they were the real scientists: lack of credentials becoming an advantage. Despite their anger, Heuvelmans and Sanderson rightly deserve credit for popularizing cryptozoology, but another writer did something similar.

In a series of books and articles in the 1940s and 1950s Willy Ley (1906–69) addressed the topic of strange and mysterious animals. A German paleontologist turned aerospace engineer, Ley counted Werner von Braun a colleague. Unlike Braun, Ley refused to put his expertise to work for the Nazis and so left the country in 1935 and came to America to pursue a career (like Ivan Sanderson and Bernard Heuvelmans) as a science writer. He produced a number of pioneering books and articles on space flight and rocket design and became something of a pop culture figure through his extensive writing and as a columnist for *Galaxy* magazine. He also dabbled in writings on unusual animal life well within the parameters of what would be called cryptozoology, but what he called "romantic zoology." In 1949 he published "Do Prehistoric Monsters Still Exist?" in *Modern Mechanix*. He discussed the legend of Mokele Mbembe and began by stating, "Dinosaurs may still roam the unexplored jungles of Africa!"[58] He produced a number of books on the subject, including *The Lungfish and the Unicorn* (1941) and *Dragons in Amber* (1951). In *Salamanders and Other Wonders* (1955) he discussed the

Abominable Snowman. His discussion of ABSMs, after a quick survey of the historical texts on the subject, arrives at an explanation. If "a near-human and very primitive race" somehow existed in the mountains of Tibet, he said, "its survivors would fit the description of the Yeti perfectly."[59] In an updated version in 1959, Ley explained how a sighting of a similar creature by Russian surveyors and mining engineers in 1957 prompted the Soviet government to mount an expedition to the Tadzhik Republic to search for it. That expedition—which Ivan Sanderson referred to in his diatribe—included the historian Boris Porshnev, who would become the primary exponent of the Neanderthal survival hypothesis and a friend of Grover Krantz. Ley took *Gigantopithecus* into consideration, even *Pithecanthropus*, and argued some specialized hominid closer to humans than to apes lived at the top of the world. Showing a restraint often lacking in the field, he finished by saying "it may be wise to wait for more fossil material before a verdict [concerning the creature's origins] is attempted."[60]

Ley collected reams of source material on pygmies, giants, primates, and anomalous animals as reference for his writing, including copying out entire chapters from Bernard Heuvelmans's *On the Track of Unknown Animals*. These included material on medieval monstrous races and legends of "little people" of Africa. Ley seemed especially interested in stories of the Agogwe, a supposed group of diminutive humanlike creatures said to inhabit the forests of West Africa and who might turn out to be Australopithecines. He engaged in extensive correspondence with professional scientists and the general public alike over science issues. His zoological writings brought him communications on weird animals. One correspondent, British archaeologist P. E. Cleator, wrote to tell Ley he had just returned from a trip to South Africa, where he had visited the office of Raymond Dart who, in 1925, made the breakthrough discovery of the Taung Child he named *Australopithecus africanus*. Cleator included in his letter a newspaper clipping on the Loch Ness Monster which, he said, appeared "in the news again."[61] Ley's work also appealed to scholars. In 1951 George Sarton, editor of the influential journal *Isis* and one of the founding fathers of the modern field of the history of science, wrote to congratulate Ley on the publication of *Lungfish, Dodo and the Unicorn*. *Isis* ran a glowing review of the book by evolutionary botanist Conway Zirkle. Sarton said he enjoyed reading the book so much that "I feel moved to express my thanks" for Ley's having written it. He then went on to relate to Ley a story of dinosaurs in Africa. So pervasive was the belief that dinosaurs still roamed the region it seems the governor general of the Congo put out an edict during

World War I requiring any dinosaurs traveling at night to carry warning lights. It had to be done, Sarton said jokingly, in the interest of public safety.[62] Ivan Sanderson, too, admired Ley, once telling him his work "enchanted" him "as I have always been with everything you write."[63]

## Conclusion

Along with the pioneers of cryptozoology, a group of academic scientists and amateur naturalists and adventurers involved themselves in the search for the humanoid cryptid known as the Yeti. This creature helped launch the search for anomalous primates into the public eye and paved the way for interest in the Sasquatch and other similar forms. Unlike the common view of scientists uninterested in such a pursuit, and even actively discouraging it, the story of the hunt for the Yeti reveals a more complex situation. Some anthropologists and other scientists took to the subject with enthusiasm to the point of actively engaging in the hunt. The 1950s and 1960s can be seen as a golden era of cooperation between amateurs and academics in the search for manlike monsters.

# Chapter 2

# The Snowmen

*It was not until the world was thoroughly explored that science banished monstrous human races to the realm of fancy, but those whom science has not yet reached may still create counterparts to the ancient tales of men with physical peculiarities.*

Eugene S. McCartney, 1941[1]

Grover Krantz's interest in anomalous primates like the Sasquatch began, as it did for so many, with the excitement over the Yeti. Romantic tales of hairy, manlike "Snowmen" in the frozen wastes of Nepal and Tibet fired the imagination. Big and hairy, the Yeti lived in a remote part of the world, the locals feared it, and it reportedly behaved like, well, a monster. It conveniently occupied a space, both temporally and physically, at the crossroads of the worlds of East and West, and so it took its place alongside other oriental "mysteries." While reports of and references to this creature may appear in Tibetan legends several millennia ago, mentions of such a creature in Western literature appeared in an 1832 report by Brian Hodgson, an Englishman living in Nepal. In the later nineteenth century another Englishman, W. A. Waddell, reported on the creature.[2] Historically important, these works did little to generate interest in the topic outside of the mountaineering community. Author Peter Bishop argues that the Yeti became popular because Western enthusiasts of the East made the region of Nepal and Tibet into a fantasy land in their minds. The actual people meant little, while Western authors turned the landscape into a paradise of spiritual awakening for themselves. Often described as a supernatural thing, the Yeti outdid the Nepalese and Tibetan peoples as a more appropriate denizen of the fantasy.[3] In the end neither books nor adulterated western pseudo-religious philosophy brought the world's attention to anomalous primates, a name did.

The commonly used Western word, Yeti, likely comes from the Sherpa word *meh-the*, or some variant, which often interprets roughly as, "that thing there." The Sherpa have numerous names for the creature, but Yeti is the one that stuck. The origin of the most popular name for the creature resulted from mangling a local language to create something acceptable to the Western tongue and imagination. In 1920 mountaineer C. K. Howard-Bury led an expedition in Nepal during which he saw several strange, manlike creatures at a distance. As soon as he could he cabled this incident back to the nearest civilization, and journalist Henry Newman of the Calcutta *Statesman* intercepted it. Howard-Bury mispronounced and mistranslated one of the Sherpa names for the creature, and Newman in turn garbled it and wound up with an English translation: Abominable Snowman.[4] This striking name caught people's attention and, more than any of the sightings or stories up to this point, made the creature a topic of interest outside the Himalayas. With this catchy but inaccurate label, the creature went from being a quaint local legend to an international phenomenon. The same type of makeover and jump to celebrity status occurred later in the century, when Sasquatch became Bigfoot.

## The Shipton Photos

Though the creature now had a recognizable, star quality name, no physical evidence of its existence had reached Westerners. That changed in 1951 with the publication of photographs of footprints in the snow of Nepal, taken by an expedition led by well known mountaineer Eric Shipton (1907–77). Born in Ceylon of British parents, Shipton set off in 1951 into the Himalayas on another in a series of Mount Everest related trips. Everest became of interest to mountaineers in the 1850s and 1860s after the British "Great Trigonometric Survey" determined it to be the tallest spot on earth, at over 29,000 feet. It took its name from a former director of the Royal Geographic Society when the name the Sherpa people used (commonly *Chomolungma*) proved difficult to determine because the political climate at the time kept British researchers out of the immediate area.[5]

Squeezed between India and Tibet, Nepal occupied a precarious geopolitical position with China looming over them all. The close proximity of the British Empire to China caused considerable tension. Both Nepal and Tibet struggled to remain independent in the shadow of these giants. In 1949 Communist revolutionaries under

Mao Zedong took control of China and the following year China invaded Tibet and claimed it. The Buddhist religion of Tibet did not mix well with Communism, so the Chinese worked to stamp it out. Tens of thousands of Tibetans were imprisoned, executed, and otherwise abused as a Chinese army of occupation moved in and brutally crushed dissent.[6]

Eric Shipton had been planning an Everest assault since at least 1945, but the war hampered doing anything more than planning. Initial ideas about a joint British and American operation were quashed by the British Alpine Club, which sponsored the climb. The club wanted a smaller party. The Indian government also hesitated to allow any climbing before 1947 because of the political turmoil of the postwar world. The British did not have a monopoly on Himalayan mountain climbing. American, Swiss, and Australian groups all prepared to go to the roof of the world. Shipton also had ideas about, and experience with, the Snowman. He claimed to have "come across many of these tracks in various parts of the Himalaya and Karakorum."[7] In 1945 an article stated that "Eric Shipton once climbed 28,000 ft up Mount Everest [before the war] and among other thrilling stories...is his report of the footprints of the 'Abominable Snowman'— huge footprints which he followed across the snow."[8] In 1951, in the Gauri Sankar area near the Tibet border, Shipton and his team had encountered and photographed a long line of strange footprints in the snow. The pictures seemed to show the track of a two-footed creature walking along at an unusually high elevation in the forbidding and freezing landscape. The first images seen in the West associated with the Snowman legend, they drew worldwide attention.[9] This attention intensified two years later, on May 29, 1953, when members of Shipton's team, Edmund Hillary and Sherpa guide Sardar Tenzing Norgay, became the first to climb Mount Everest.

The publication of the Shipton photos created a sensation not just with the public, but among scientists of the day. Soon, Shipton contacted Yale University ornithologist Dillon Ripley for advice. Sidney Dillon Ripley (1913–2001) hailed from a wealthy railroading family. Following her divorce, Ripley's mother sent him and his sister on a six-week sojourn to India and Tibet during which the young Ripley developed a lifelong attachment to Asia and its birds. During World War II he joined the American OSS (Office of Strategic Services, which after the war became the Central Intelligence Agency) and because of his prior experiences and expertise was posted to the India-Burma theater of operations as a civilian intelligence operative. His job was to organize field agents with codenames like LOLLIPOP and operations called JUKEBOX and BALONE.[10] He also arranged for submarines to pick up and deposit agents across the Far East. Ripley also established himself as a world renowned expert on the birds of Asia, publishing several books on the subject, including *Birds of Bhutan* (1996) with Biswamoy Biswas (1923–94) the Indian ornithologist who would later serve on the *Daily Mail* Expedition to find the Yeti.[11] He regularly consulted on ornithological questions for those going into Nepal and eventually became secretary of the Smithsonian Institution, 1964 to 1984, playing a brief role in the story of the Minnesota Iceman.

In 1951, immediately after the Shipton photos came out, Dillon Ripley received correspondence from a Russian émigré, Boris Lissanevitch, then living in Nepal, who said he wanted to mount an expedition to catch the Yeti. An unusual character, Lissanevitch (1905–85) began as a ballet dancer in his youth, leaving his homeland after the 1917 revolution. He wound up in India, then Nepal, opening a nightclub and settling down enthusiastically into the role of local fixer. Gossip held that nothing could get done in Nepal without the help of "Boris of Xathmandu." Boris seems to have known everyone who came to Nepal, including the Yeti hunters. He also seems to have been accused by everyone of being a spy for everyone else. Boris consulted Ripley about the idea he had of catching the creature with nets he had imported into the country just for this purpose.[12]

After finding the footprints, Eric Shipton also wanted the advice of the scientist on how to further the search for the Yeti. Chastened by Boris the Russian and his plan of throwing a net over the Yeti, Ripley wanted to be careful.[13] He told Shipton: "Dr. Coon has pointed out" that one need not kill or capture a Yeti to prove its existence. He should look for evidence like a cave dwelling or related artifacts. Ultimately, Ripley cautioned the mountain climber against certain search behavior and what to avoid, and did so with prophetic words. "This whole subject," Ripley told Shipton, "is one open to ridicule by scientists," and "that a little detective work on the quiet of this sort would perhaps be the first step."[14] Ripley instinctively knew that rushing in and making claims before solid evidence turned up, or making the search a circus would only serve to push the scientific community away from the Yeti and make the search a joke. He also cautioned that even discussions should be done away from the public eye. Wise advice, but advice few followed in the coming decades. The search for anomalous primates suffered its greatest setbacks largely due to the overly enthusiastic pronouncements and actions taken by amateur monster hunters. Scientists, whose work is normally done slowly, methodically, and quietly, could be scared off the monster trail like squirrels inching toward a peanut startled by a loud noise. They rightly feared their careers, built up over years of effort and formal schooling, could be ruined with one stroke of the absurdist brush, which often rightly or wrongly painted the picture of crypto-zoological research.

The standard narrative of the crackpot versus egghead story is that the eggheads refused to even consider the existence of such creatures: from the beginning they dismissed the entire idea with a contemptu-ous wave of their collective hands. This is just not the case. Despite the media frenzy, scientists in the United Kingdom, United States, India, and Russia wanted to begin investigating the Yeti immedi-ately after Shipton published his famous photos. The "Dr. Coon" Dillon Ripley mentioned to Eric Shipton turned out to be Carleton Stevens Coon (1904–81), a Harvard trained anthropologist with a specialty in the anthropology of North Africa and the Middle East. Coon already had a reputation as the author of a series of controver-sial books on human origins and behavior, race, and intelligence.[15] These included *The Races of Europe* (1939), and later would appear *The Origin of the Races* (1962), which held that while many ethnic groups around the world had excellent characteristics and attributes, ultimately Europeans led the way. He also had been looking into the Yeti question on his own. In *The Story of Man* (1954) he would make brief reference to the Abominable Snowman and the Shipton footprints. Coon suggested the creature might be a descendent of *Gigantopithecus* and speculated that "man may not be the only erect bipedal primate to have survived the Pleistocene period."[16]

From the 1920s through the 1950s Coon studied the ethnog-raphy of local populations from Morocco to Ethiopia and Iraq. He worked on Stone Age sites as well as studied modern peoples. Like Dillon Ripley, Carleton Coon had served with the OSS during the war, contributing to the Allied invasion of North Africa. Despite the fact he had a professorship at Harvard, Coon had initially tried to join the navy in 1941 but his obesity kept him out. He went on a reducing plan and then began to pester the government to hire him for intelligence work in North Africa and the Middle East. They grudgingly relented and attached him to the OSS as an agent to the American Legation in Tunisia. There he recruited the local Riffian people to fight with the Allies against the Nazis and Fascists, received covert military training, and performed combat missions, often in British Army disguise. In 1942, while operating with local elements, a German bombing raid caught him in the open. He managed to duck into a house but it collapsed around him, and a heavy roofing tile hit him on the head. Coon managed to escape but some time later began to slur his words, suffer from headaches, act irritably and violently, and lose his vision. A medical examiner evacuated him back to Boston for brain surgery.[17]

After a recovery period, Coon wanted to get back into the fray but the OSS thought differently. He managed to persuade the U.S. Army of his usefulness, and in 1943 he was commissioned as a major and went back to duty. Though he insisted he was fit, his work suffered, and his symptoms continued. By 1944 the army decided he no longer had the capacity to perform his duties, as his head injury rendered him "incapacitated for any active service." He underwent a full medical exam and went before a retirement board. The medical records from the CIA show that Coon had suffered from nervous problems since the mid-1920s when he had incurred an initial head injury. The wound in Tunisia exacerbated that condition. He now suffered from hysterical psychoneurosis and severe anxiety of a permanent type compounded by his wife's divorcing him in the middle of it all.[18] The examining board released him from his commission.[19]

With his wartime service at an end, Coon went back to his academic pursuits. As an anthropologist focused on human evolution and behavior, legends of the Yeti fascinated him. When the Shipton photos appeared with their apparent proof of the existence of the Snowman, he began investigating seriously. In discussions with Dillon Ripley in 1953, Coon referred to the Yeti as "our dear friend," and exclaimed, "Up *Gigantopithecus!*" He, like Ripley, feared too much publicity would make it impossible for any real anthropological work to be done. After the initial attention died down a bit, he breathed a sigh of relief that "I think we have passed the fantasy stage and are ready to do some serious thinking and planning" about going to find the creature.[20]

Unfortunately, Coon spoke too soon about the "fantasy stage" being past. As he and Ripley were hoping the situation had reached a point where they could get to work on the problem, the next phase of ballyhoo started. In 1953, on his way to cover the expedition that would eventually reach the top of Mount Everest, journalist Ralph Izzard had a chance encounter with team leader, Major John Hunt, who believed in the reality of the Yeti. Hunt suggested to Izzard that an expedition specifically to find the Yeti should be mounted. Izzard thought it a good idea and proposed it to his superiors at the *Daily Mail* back in London, who saw a great way to capitalize on the Everest craze.[21] With that, the *Daily Mail* Expedition was born. Ripley's and Coon's hopes for proper inquiry into the Yeti were dashed as the major expedition phase of the anomalous primate hunt began. Modeled on the typical large-scale mountain climbing expedition model, a group of mountaineers, scientists (including Biswamoy Biswas), and hundreds of porters headed off to find the Snowman. They tramped about the Himalayas for 15 weeks, finding little more than some inconclusive tracks, a few strands of hair, and animal droppings. The legacy of the *Daily Mail* Expedition, however, stems less from what it found or did than from the reactions it produced in other monster enthusiasts. The waves it generated reached from Nepal all the way back to the New World.

## The CIA and the Snowman

The *Daily Mail* Expedition aroused the interests of many amateurs. One of the more significant was Thomas Baker "Tom" Slick (1916–62). An eccentric, enigmatic Texas oil millionaire, Slick had an interest in monsters, and he financed a number of expeditions looking for the Yeti and Sasquatch. In fact, Slick's passion and largesse helped many of early founders of anomalous primate research—including René Dahinden and Peter Byrne—pursue their work. Slick's largess would also be a source of frustration at times.[22]

Slick had been drawn to Asia and its mysteries since childhood.[23] He traveled to the region several times and heard legends and rumors of the Yeti. Besides monsters, international politics fascinated Slick. Though born in Texas and holding a lifelong attachment to the Lone Star State, he had a clichéd Eastern elite establishment upbringing, including Yale University and membership in the Skull and Bones fraternity. He moved in the same mid-twentieth century circles that produced future CIA director Allan Dulles, conservative publishing mogul Henry Luce, and other proponents of the American Century ideal—including both Presidents Bush. He traveled through Nazi Germany and Soviet Russia in the 1930s, later developing a taste for Nazi memorabilia. Politically a moderate, conservative Republican, Slick's ideas were given a quirky temper by an interest in global peace studies, and by an almost utopian, liberal notion of bringing the world together through the expansion of the human mind rather than the brute force so many of his peers advocated. While it is difficult to determine anything for sure about Slick's *weltanschauung*, he may have rebelled against the straightlaced formalism of mid-century conservative American exceptionalism prevalent in his class, and have reached for something higher. He articulated such sentiments in *Permanent Peace* (1958).[24] Slick argued for the United States to join with other nations, including those of the Communist bloc, to create a kind of world police force to help ensure peace and stifle war. In addition, the United States would have to give up some sovereignty in global matters, even some of its overseas possessions. It would have to, along with all other nations who had them, reduce and eventually eliminate its nuclear arsenal.[25] To further this end, Slick sponsored the 1961 *Conference to Plan a Strategy for Peace*, held in New York at Columbia University. The discussions included such notions as that "the government of the People's Republic of China [then considered an enemy of the Western powers] be drawn into fuller participation in the international community."[26] He also founded a series of metaphysical institutes, including the *Mind Science Foundation* of San Antonio, Texas. Along with these interests he dreamed of monsters.

Stirred by the *Daily Mail* Expedition, Slick contacted a number of individuals about mounting an expedition of his own. In 1956 he made an initial visit to the area to get the lay of the land and talk with locals. While speaking with Tenzing Norgay, of Everest fame, he learned of Peter Byrne, also trying to get an expedition going. Born in Ireland, Byrne had been a tea grower in India who, like so many other Europeans and Americans, became obsessed with the Yeti legend and was determined to find one. He left Asia for Australia, where he worked as a journalist and at drumming up financial support for a Yeti expedition (Byrne proved to be one of the most adept of the monster hunters at getting financing for the quest). In 1956 he returned to Nepal, and he and Slick eventually met and agreed to work together. Biographer Loren Colman is convinced that Slick and Byrne probably worked for the CIA in some capacity.[27] Slick's niece, Catherine Nixon Cooke, dismisses such claims, saying there is nothing referring to the CIA in Slick's papers.[28]

As Tom Slick assembled his expedition, the hugely popular *Life* magazine stepped forward as a potential partner. The magazine offered Slick $25,000 up front for the rights to any photos he took of the monster. They also separately approached Carleton Coon for his thoughts on organizing an expedition he would lead, and at the same time asked him to keep an eye on Slick and his efforts. Coon agreed, and the magazine sent him to Katmandu. In his autobiographies, *A North Africa Story* (1980) and *Adventures and Discoveries* (1981), Coon mentions little about this aspect of his work other than to say that "my relationship with *Life* was felicitous from start to finish."[29] He referred to Tom Slick, the object of his surveillance, as "a very nice guy."[30] The extant correspondence from this period, however, show Coon held a different attitude in private toward the Yeti and Slick.

Carleton Coon took *Life's* offer quite seriously and began working out ideas and questions about how to mount an expedition to find a Snowman, which in his notes he refers to as the "High Altitude Project." He made it clear that any such undertaking should be to photograph the creature only. If they came across a carcass, they would take it, or should an emergency situation arise in which a creature had to be shot for defensive purposes, they would do so, but the primary aim had to be picture taking. He laid out details of how many people would be needed and their roles: he suggested a minimum of six and maximum of nine. The party would include his wife, and his colleague, Fred Ulmer of the Philadelphia Zoo, because of his expertise in primate anatomy. Coon worked out a budget of $100,000 and suggested CBS television should send a camera operator so the expedition could be filmed. As to what to do with material collected, "Any physical remains of snowmen or of artifacts which they may produce shall be the property of the University [of Pennsylvania] Museum." In the unlikely case of capturing one alive, "the Philadelphia Zoo shall have first refusal."[31] While he thought it a dubious idea, he did give consideration to capturing a live specimen. If done properly and with plenty of preparation it could be pulled off, he thought. However, weapons would be needed and it was historically difficult to get permits to carry into Nepal the type of firearms that would bring a Snowman down.[32]

Coon's interest in human behavior and evolution drove his interest in the Yeti. To study a Yeti, he said, "would be of incalculable help to those of us who are trying to find the origins of human behavior."[33] Coon believed the modern human ethnic groups, the races, had all formed separately out of *Homo erectus* populations and then later evolved into modern *Homo sapiens* as subspecies. This concept would, in a later (different, subtler, and more complex) and less race-centered incarnation, be known as multiregionalism (While not referring to it by this name, Grover Krantz would employ a hypothesis based upon Coon's work). Coon had trained at Harvard in the 1920s under anthropological legend Earnest A. Hooton. Coon, as did Hooton, used a technique known as *typology* to study human ethnic groups. This approach focused on the supposed differences in human morphology to distinguish one ethnic group from another, and to rank them according to supposed levels of superiority and inferiority. Instruction and research at Harvard at that time took more of a hard-anthropology approach by centering on physical attributes.

In direct contrast, Columbia University's soft-anthropology approach made famous by Franz Boaz, saw culture and society as the way to explain ethnic difference and rejected the entire concept of "race" as a useful scientific term. Coon generated resentment among his colleagues not only for his views on race, but for not taking a public stand on the social implications of his work, despite segregationists and racist forces in the southern United States during the turbulent civil-rights era of the 1950s making use of it to give scientific authority to their cause.[34]

After careful consideration of the materials he had available at the time, Coon wrote in his notes that the Snowman was likely "a bipedal, cold-altitude adapted oversized form of the gibbon-siamang genus or a related species."[35] As he sat at his desk mulling over his thoughts on what might happen on a Yeti hunt, Coon began to doodle on his notes. While convinced the creature was real, he also acknowledged the level of oddness and inherent humor such an expedition entailed. He drew a little cartoon of himself climbing a mountain at the top of which a Yeti waited with a club to bonk him on the head while flying saucers hovered nearby. The hunt would be strange enough with the inclusion of the alleged espionage element without the addition of UFOs.[36]

The intelligence angle of the search for the Yeti is a complex and often contradictory one. In 1950 and again in 1951 Coon had been temporarily appointed as "an unclassified scientific consultant" to the CIA at a pay rate of $35 a day. This position allowed him to provide "occasional overall guidance on [words blacked out] intelligence problems, particularly on Arab states."[37] He supplied this service during the research tour of North Africa he made for a book project. CIA records provided regarding Coon's career do not go beyond these dates except for the records of a disability claim in the late 1970s. There seems to be no evidence that Coon worked, at least in any official paid position, for the CIA during the Tom Slick expedition period.[38] *Life*'s founder and nominal Coon sponsor, Henry Luce, however, had a reputation as an anti-Communist. They all seemed to be watching each other. *Life* may have simply wanted to make sure Slick did not scoop them on discovering the Yeti or sell his pictures to a competing magazine, or Luce may have worried over Slick's peace activism, which included trying to establish contacts in Communist China over the Tibet issue.

Henry Luce (1898–1967) was the lord and master of *Life* magazine and a large publishing empire. Born in China, Luce had a lifelong attachment to the country and had strongly opposed the

Communists in their bid to take control. Luce had organized relief missions and supported General Chaing Kai-shek, putting him and his wife on the cover of the magazine numerous times. In 1941 he personally authored the *Life* article titled "The American Century," which helped persuade an entire generation to believe in America's destiny to rule the world. Slick's view of the world, despite sharing a social lineage with Luce, did not fit with the American Century ideal. With his CIA connections, Coon fit better, so it is not surprising that *Life* went with him as a Yeti expedition leader. Just as sides were forming between Slick and *Life* in the race to find a Yeti, the two rivals suddenly joined forces, or at least came to some type of understanding.[39] In January 1957, Jim Greenfield, *Life's* New Delhi bureau chief met in Darjeeling with Peter Byrne, Slick's girlfriend/ assistant Cathy Mclean, and Carleton Coon to discuss the Slick expedition.[40] Mclean had met with Byrne the year before to size Byrne up as a potential expedition member.[41] What they discussed in the Coon interview remains unknown, but following this meeting Carleton Coon now worked with the Slick expedition and the *Life* expedition quietly disappeared. Joining forces would be useful, as many difficulties cropped up in hunting monsters in Nepal.

One constant obstacle to the monster hunters took the form of seemingly endless reams of official paperwork and permits needed for access into restricted areas of Nepal, especially anything involving Westerners and close proximity to the Chinese border. The frustration grew for men of action like Slick and Byrne, who wanted to get on with it and not be held back by bureaucratic red tape. The official backing of some reputable academic institution would make things go more smoothly.[42] Luckily, Slick sat on the board of the San Antonio Zoological Society of Texas. In Carleton Coon, he also now had a reputable scientist on the project as well. Slick apparently did not know Coon had been watching his operation for *Life*. Regardless of who was watching whom, the two rival expeditions had combined into one: at least for the moment.

Evidence that Carleton Coon's interest in Nepal went beyond spying opportunities comes from the considerable amount of preparation he did in anticipation of the Slick expedition. He created a compendium of past sightings and wrote up notes on primate behavior as well as questions to ask concerning the many details of such an undertaking. By the next year, however, the mercurial Slick decided to pare the operation down so that only he and Byrne and a small team of close confidants would actually travel to Nepal. Why he did this is unclear. It is possible that he found out about Coon's spying and wanted him out of the loop. As a result, Carleton Coon remained behind as only a scientific consultant for the Slick expedition. One who would go was expedition cosponsor Kirk Johnson, director of the Forth Worth [Texas] National Bank, who had no zoological expertise.[43] Unlike the *Daily Mail* Expedition, which had made at least a passing gesture at scientific legitimacy, there would be no trained scientists on the Slick field team. Academics would stay home in the United Kingdom and United States and wait for samples to be sent to them for analysis. Not having any proper scientists in the field to determine what evidence had value and what did not would be the root of much bad feeling and frustration and would contribute to the scientific world's rejection of the entire notion of monster research.

## The Yeti's Hand

Like all the various monster expeditions before and since, the Slick project wandered around the region finding a few bits and pieces here and there: a few possible footprints, some hair samples, an apparent Yeti dung sample, lots of local stories, and not much else. The only "discovery" came in the form of materials already in someone else's possession and known to other Yeti hunters. At the remote Buddhist temple at Pangboche they encountered the Yeti's hand. The *Daily Mail* Expedition had already found the monastery held a repository of Yeti relics. They had examined the "scalp" monks at the monastery claimed had come from a Yeti. The monks allowed the explorers to carefully remove a few hairs from the scalp, which were sent back to England for examination. Anatomist Frederick Wood Jones (1879–1954) examined the hairs (just before he passed away) and said they were probably from a local goat called a *serow*.[44] In addition to the scalp, the Slick team wanted to see the Yeti's hand.

The temple these interloping foreigners had traveled so far to see had a long and revered history in the Sherpa community. The Pangboche location already served as a population center when the early Sherpa people arrived in the region in the mid-1400s. Years later, according to Sherpa/Buddhist tradition, a Lama named Sangwa Dorje walked there from India and settled in a cave to meditate and take up the life of a religious recluse. Friendly Yetis are said to have helpfully brought him food and water, allowing him to survive and keep praying. When one of the Yetis died, Sangwa kept his scalp as a holy relic. The local people, so awed by Sangwa Dorje's religious devotion, begged him to build a *gompa*, or temple. Returning to his meditations, he contemplated what to do. Next, a holy statue "flew from India to his hermitage" and instructed Sangwa how to proceed.[45] He built the temple, consecrated in 1667, and installed the statue and the scalp. For generations the location was a center of both political and religious activities, until political power eventually shifted elsewhere and the Pangboche temple became mostly a religious site. Just when the Yeti hand arrived is unclear, but the monks held it in high religious esteem. The Sherpa people saw the relics, not as rude curiosities, but as tributes to the Yetis that assisted Lama Sangwa Dorje in the sacred creation of the temple. The Slick expedition did not see or treat the relics as religious icons; they wanted to get the hand for scientific analysis. The monks allowed team members to view the hand, but steadfastly refused to allow it to be taken away.[46]

Having heard of the existence of the hand, British primatologist William Charles Osman-Hill encouraged Slick and Byrne to get a piece of it to study. Byrne ingratiated himself with the monks over time, waiting for his chance. He even donated a special fabric-lined case in which to store the relic. In 1959, when the moment was right Byrne stealthily removed several finger bones and substituted human bones. He then carefully wrapped the hand so that nothing would seem amiss.[47]

Primatologist William Charles Osman-Hill (1901–1975) was another academic with an interest in monsters. He had a long distinguished career at the Yerkes Primate Research facility and a number of other university laboratories. During the 1930s and 1940s, he spent 15 years in Ceylon (Sri Lanka) where he conducted extensive research into the zoology of the region. He also kept a large private menagerie of birds and animals which he donated to the London Zoo.[48] In 1945 he investigated the legend of the Nittaewo, the fabled little people of Ceylon. He concluded the Nittaewo might well be relic populations of the fossil hominid, *Pithecanthropus*.[49]

Upon examining the bones of the Pangboche hand, Osman-Hill found himself puzzled. First, he said they looked human then changed his mind, suggesting they had Neanderthal characters. His work on the Nittaewo had predisposed him to the "relic" population theory, which proved a durable and popular way of explaining anomalous primates. For some it seemed most likely that monster legends referred to small isolated groups of early hominids that somehow managed to survive long after the bulk of their species had gone extinct. There are examples of small groups of modern humans who had found themselves cut off from the rest of humanity and whose societies seemed relics of the Stone Age past. If these groups of *Homo sapiens* had managed to go undetected for so long, the reasoning went, it may have happened with earlier groups of proto–*Homo sapiens* as well. Ivan Sanderson, who also championed the relic population concept, claimed Osman-Hill was "shook up" by the Yeti droppings because they seemed to suggest an unknown primate had produced them.[50]

Tom Slick was not going to take scientists like Osmond-Hill with him to Nepal, but he had them at home ready to examine what he found and deigned to send them. They all had expertise in Asian flora and fauna as well as other connections. In addition to Carleton Coon and Osman-Hill, the list of scientific consultants included Bernard Heuvelmans, Ivan Sanderson, and journalist Ralph Izzard. A second tier of scientists, whose only direct contact to the expedition came through Coon and New World archaeologist and anthropologist George Agogino (1921–2000), of the State University of South Dakota, received materials for double checking and specialist tests: all of them initially jumped at the chance to get involved.

With Osmond-Hill in England, Tom Slick's pivotal North American consultant was Agogino, familiar with Native American Sasquatch folklore as well as Yeti legends. Agogino acted as a kind of central exchange, getting materials from the Slick team and then distributing them to the others for analysis. Like Dillon Ripley and Carleton Coon, Agogino had served with the OSS in Asia during the war—many of the scientists investigating anomalous primates during this early phase had military intelligence as well as Asia in common. The Soviet government in Moscow noted the intelligence connection with Slick's operation and saw it as "a diabolical anthropological maneuver aimed at the subversion of Communist China." In classic cold war rhetoric the Russian newspaper *Izvestia* said the Americans only searched for the Yeti as a cover to spy on and gather intelligence about the Nepalese-Tibetan border. The Chinese government compared the Slick expedition to others then searching for the biblical relic of Noah's Ark along the Russo-Turkish border.[51]

On October 4, 1957, the Soviets launched Sputnik and the world turned its attention to the heavens. The International Geophysical Year also began in 1957, a worldwide operation with dozens of countries engaging in research activities on various earth sciences, especially in the polar and frozen zones of the world. While this all went on, the Russians also looked for a Snowman. Not to be left behind with a monster gap, Russian scientists began nosing around the Pamir Mountains northwest of Nepal and Tibet, along the Chinese border. Russian investigators like Sergei Obruchev and Boris Porshnev "collected all available material in the Soviet Union, China and Mongolia" on the Yeti.[52] As will be explained later, the Russians, officially at least, seemed to go back and forth during this period. In February of 1958 an *Izvestia* article said scientists did not believe the creature existed, despite the fact that others, like A. G. Pronin, claimed to have seen a Snowman in the Pamirs. Pronin had been negotiating the Fedchenko Glacier when he reportedly saw the creature.[53] By November of that year another article showed scientific optimism about making a discovery that would stun the world.[54] Further consideration, however, prompted reporters to say the "myth" of the creature had been "punctured."[55] Eventually, a cadre of Soviet scientists focused considerable effort in the search for Eastern anomalous primates and later joined forces with their Western capitalist colleagues.

In the late 1950s and early 1960s, the mountains of Central Asia saw quite a bit of monster hunting activity. Throughout this period political tensions grew. In March of 1959 an uprising in the Tibetan capital of Lhasa brought further Chinese repression and throngs of Tibetans, including the Dalai Lama, fled to India. The Chinese had no interest in having Westerners—especially suspect monster hunters—watching any of this up close.

### George Agogino

When he joined the Slick project, George Agogino held a position teaching anthropology and archaeology in South Dakota. Like so many academics in this story, Agogino was at first unsure as to whether the Yeti existed, but he grew intrigued by the idea and felt it a reasonable scientific pursuit to find out once and for all. He accepted Slick's request to act as a consultant. He also quietly contacted Carleton Coon for advice on how to deal with Slick. Agogino and Coon began to correspond regularly and had an entire discourse on the proceedings outside the official Slick expedition correspondence track. They did this for a number of reasons. First, while Coon was in the Slick camp, he had also been wooed by *Life* magazine to be part of their Yeti project and to spy on Slick.[56] Secondly, Coon (and then Agogino) had suspicions about Slick, his methods, and motivations. Having been enthusiastic about going on the Yeti expedition, Coon remained upset with Slick for being demoted off the field team at the last minute. Agogino met with Coon at his Harvard office for lunch just after Christmas of 1958 and discussed how to deal with the wealthy Texan. Coon told Agogino to watch his step, to which Agogino replied, "You can be sure I will be most careful in my dealings with him." He also lamented how some of their colleagues discounted the idea of a Yeti, unwilling to look into it. Determined, he told Coon, "Somebody must do this work or the situation will never be cleared up. It just makes me angry we must deal with individuals like Tom Slick."[57] Agogino appreciated Slick's commitment, and his money, but was troubled by his and his mountaineer associates' lack of scientific training and apparent focus on adventure rather than research. He liked to refer to Slick's entourage as "the Slick Mob."[58]

In 1959, as the Dalai Lama was fleeing his home, Agogino began to receive field samples of supposed Yeti dung from Nepal. He sent them off to Heuvelmans, Coon, Osman-Hill, and Izzard. They all (except for Izzard, who had no such analytical expertise or facilities) concluded the droppings did not come from a primate. When Slick sent Agogino a small black and white photo of the Pangboche hand for confidential inspection Agogino sent a copy to an unimpressed Carleton Coon.[59] That no trained zoologists accompanied the team in the field to determine useful from useless materials, and no one with training in technical photography to record finds, quickly showed its drawbacks. The mountaineers and adventurers who made up the team had no experience making such decisions or producing a proper photographic record. Had Coon gone along, as he thought he should have, an enormous amount of time and resources would not have been wasted. This concerned Agogino more than ever. Instead of preparing for important details, the Slick people had made preparations for situations they had no experience with and which Agogino felt unlikely to occur. He told Coon, "The Slick Mob is resorting to live traps and power weapons with drugs in the payload instead of lead."[60] During his trip, Slick encountered "Boris of Xanthmandu." Slick utilized Lissanevitch's services to smooth the way for his expedition and, in return, presented the Russian with one of the "power weapons" Agogino referred to. Slick had inquired about tranquilizer guns with Cornell University as well as the American Museum of Natural History. It was an odd contraption Lissanevitch's biographer—and Yeti debunker—Michel Peissel called an "Alka Seltzer gun," a kind of tranquilizer projector which had both a sleeping agent and one to wake the creature back up. Lissanevitch feared lending it to anyone or even touching the thing himself.[61]

George Agogino thought it the wildest fantasy to try to catch a Yeti alive and bring it back home like King Kong. As did Coon, Agogino thought the emphasis on catching a Yeti ill-advised and did not like the way the Pangboche hand had been treated. The field team also sent back photos and X-rays of another artifact called the "Makalu Hand." All the scientists who looked at the photos concluded dejectedly that it was from either a wolf or a snow leopard, not a legendary hairy monster. Agogino also grumbled about how much work he did for Slick without getting paid, and he did not care for the press coverage or for the increasingly circus-like atmosphere. He complained to Coon: "I certainly hope our profession can do better than that."[62] The two scientists tried to determine Slick's deeper motivations, but without success. Agogino voiced his concerns about the Slick expedition to Ralph Izzard and William Charles Osman-Hill and told Coon "they seem as confused as we" as to Slick's intentions.[63]

Despite his reservations, Agogino continued his commitment to the problem. He sent Coon a list of everything that had been collected, and sent hair samples to zoologist Charles Leone of the University of Kansas for serological analysis, and similar specimens from the Pangboche scalp to retired Rutgers University zoologist and mammal hair and blood specialist Leon Hausman (1889–1966). Hausman had tested the first "Yeti" hair samples to come out of Nepal in 1953, so Agogino went to him in 1959. Rutgers had recently opened the first major serology, or blood research, laboratory and it was hoped the new facility might crack the case of the Yeti's blood. Hausman found the 1953 sample to be several hundred years old, consistent with a religious relic, but made of hair from an animal unlikely to be the monster of Asia.[64]

By now, Agogino found himself in an awkward position. He felt the Yeti phenomenon should be investigated by scientists, and while he appreciated being on the receiving end of Slick's materials, he did not appreciate the manner in which they were acquired. He started creating a private file of his dealings with Slick "in the event that Tom Slick misquotes anyone or breaks away from me." He told Coon he insisted to Slick that none of his reports be made public until an approved final report could be produced, and that "he better not violate this rule." He also did not want to antagonize Slick for fear the stream of materials being sent would dry up should some actual evidence be found.[65]

By the spring and summer of 1959, disillusionment with the Slick project reached a peak. After careful examination of photos of the Pangboche hand, Coon told Agogino, "Peter Byrne is no better photographer than he ever was." To his anatomist's eye, the hand was clearly human not primate. Coon felt more testing on the Makalu hand a waste of time as well. Continuing to hope, however, he said, "I would like to see the X-rays if possible; they will cinch the diagnosis."[66] Agogino sent him more materials to examine, including another hair sample, a gnawed twig, and yet another fecal sample, and asked if Fred Ulmer could look at it.[67] Coon examined this material and wearily told his partner in "Snowmaniana" that it was nothing. "As for the turd," he continued, "it has little in it." Fred Ulmer examined the material but decided not to comment because "he is tired of the whole thing."[68] Agogino, too, felt the effects. "I am running out of steam and ideas," he told Coon. By then Charles Leone had completed his tests and had come up with nothing. Fred Ulmer thought the footprint cast Slick had sent a fake, though Agogino wondered how or by whom a fake would have been made in Nepal. Adolf Schultz, the director of the Anthropologisches Institut Der Universität Zürich, thought the print that of a panda.[69] Frustrated by the lack of advancement, Agogino asked, resignedly, "Where do we go from here? I have tried almost every research angle I can think of."[70] He admitted that "I started out to expose this as a hoax," but despite the lack of progress he found it "harder to do than I first suspected," but no conclusive material seemed to exist to prove anything one way or another.[71]

In addition to Asia and the intelligence community, another disturbing connection between these early monster hunters manifested in the way of a possible shared ideology. One other scientist Coon corresponded with over the Yeti was Italian statistician and demographer Corrado Gini (1884–1965). A prominent fascist theorist in the 1920s and 1930s, and close to Italian dictator Benito Mussolini, Gini shared an interest in anomalous primates with Coon and his cohorts.[72] An arch eugenicist Gini believed that nations, like individual people, went through periods of birth, youth, growth, old age then decline, and that strong nations need not apologize for their forceful expansion.[73] He also thought, like Coon, that the Snowman might shed light on human evolution and sociological structure. In the late 1930s Gini founded a society that studied genetics and eugenics—its journal was *Genus.* where in later years Ivan Sanderson and George Agogino published articles. In 2004 Russian anomalous primate researcher Dmitri Bayonov said that in 1962 Gini founded the *Comite International pour l'Etudes Humanoides Velus.* Members of this committee included William Charles Osman-Hill, Bernard Heuvelmans, John Napier, and Philip Tobias as well as John Green and René Dahinden.[74] While Coon certainly had politically incorrect ideas about race, the others had no overt fascist sympathies. They may simply have welcomed any help from the international community that supported their work —they embraced scientists from the Soviet bloc as well.

Either unaware of, or intentionally discounting, the fact that the entire enterprise seemed to be going nowhere, was riddled with spying and counter spying, and had peculiar fascist connections, Tom Slick remained enthusiastic. Writing in *Explorer's Journal* at the end of 1959 Slick said the expedition had done well and even found tracks at the 12,500-foot level. He speculated that multiple species of Snowman must prowl the earth because multiple, but consistent, descriptions of the animals encountered had come to light. He summed up, saying he believed the idea "which has been suggested before by several anthropologists," that the Yeti "might be related to the extinct ape man *Gigantopithecus.*"[75] Not long after this, one last expedition headed to Nepal, but Tom Slick did not take part.

## The Hillary-Perkins Expedition

In 1960 and 1961 Edmund (now Sir Edmund) Hillary (1919–2008) and American television wildlife personality Marlin Perkins (1905–86) went to Nepal to wrap up the Yeti mystery, they thought, once and for all.[76] Perkins, director of the Lincoln Park Zoo of Chicago, soon to be the host of the popular television show *Wild Kingdom*, had followed the Yeti saga with interest.[77] The expedition had the backing of the well known *World Book Encyclopedia*, which claimed the trip would ostensibly test the effects of high altitude on climbers. As with the Slick expedition, Communist propagandists argued the Hillary-Perkins operation was a cover for spying on the Nepalese border with China for some nefarious capitalist skullduggery.[78] Unlike the Slick expedition, the Hillary-Perkins operation brought along a brace of scientists. (Although invited, George Agogino declined to go due to previous commitments.) They made a beeline for the Pangboche monastery, where the monks allowed the team to examine the Pangboche Hand. To the media covering the story, evidence seemed to be mounting that "there is indeed still at least one known anthropoid awaiting scientific discovery in Central Asia."[79]

Hillary and Perkins found Peter Byrne's doctored hand, which the monks had yet to realize had been tampered with. Not knowing of Byrne's switcheroo, Perkins made the obvious observation of its modern human nature. Both Perkins and Hillary, already suspicious of the hand, had come at the Yeti legend from a position of skepticism. The modern aspect of the hand only seemed to confirm their position. The Yeti scalp on the other hand intrigued them. Like all the expeditions before them, the Hillary-Perkins team asked to take the scalp with them. To their surprise this time the monks said yes.

Just because the Sherpas had been taken in by Tom Slick and Peter Byrne, it did not mark them as fools. Determined to get something out

of the increasing stream of Westerners tramping through their country looking to climb mountains and chase monsters, they approached
Hillary with an intriguing offer. They would allow Hillary to take the
scalp to England and America if one of their elders, Kunzo Chumbi,
went along with it. Having been used by Westerners for their own
purposes, the monks at Pangboche now slyly used the Westerners
in return. Chumbi became a media darling and spoke to numerous
groups, using Hillary to raise money to build schools in Nepal and to
generate prestige and political power for himself back home. Another
Sherpa *mi-che* (political big man) used publicity from the expedition
to get the funds to build a much needed water system and to compete
with Kunzo Chumbi. Far from being unsophisticated Asian hillbillies
the Sherpas turned out to be shrewd politicians who quickly grasped
the notion of using to their advantage the foreign fascination with
Everest and the Yeti, and the outside view of themselves as noble
savages.[80]

Edmund Hillary took the scalp out for analysis, including visiting William Charles Osman-Hill. It did not impress anyone in its
complete form any more than it had as a few stray hairs years before.
The *Sunday Times* said it all: "The decisions of anthropologists and
other scientists in Chicago, Paris and London are unanimous—it isn't
a scalp at all." It was the old culprit: the goat called a serow.[81] While
Hillary still thought some unanswered questions lingered in the case,
the analysis of the scalp and the hand had all but sealed the fate of the
Yeti as far as interest from the scientific community.

Like the Slick expedition, the Hillary-Perkins expedition had
its share of international intrigue. A member of the team, legendary *National Geographic* magazine photographer Barry Bishop
(1932–94) who over the course of his career had taken some of
the most famous photographs of Everest expeditions, encountered
unusual outside interest. Ralph Styles, a U.S. Navy captain and member of the Defense Department, contacted Bishop upon his return
to America. Styles invited Bishop to come to Washington, D.C., for
a chat about his Hillary-Perkins adventure. There, Bishop answered
questions about the region's climate and the potential survivability of
U.S. aircrews should they go down there. U2 spy plane flights had
been coming out of Pakistan to roam over Communist bloc areas.
Should they be shot down, the pilots would be forced to eject over
the home of the Snowman. The U.S. Air Force had partly funded the
Hillary expedition and National Geographic Society director Melville
Grosvenor, a prominent Republican Party supporter, often allowed
*National Geographic* writers to have similar chats with the CIA upon

their return from overseas assignments: something that did not always sit well with the staff.[82]

## Conclusion

The details are murky, but it seems clear that at the very least a Western intelligence element existed alongside the search for the Yeti. The mixture of cloak-and-dagger spy work with anomalous primate research seems almost comical in its unfolding. It raises the question, who *was not* working for the CIA or British intelligence in this operation? Carleton Coon spied on Tom Slick, who may have been a spy himself. Coon worked with Slick, but did not trust him, nor did he care for Peter Byrne, another alleged agent. Coon told fellow spy George Agogino to watch himself around Slick, and Agogino heartily agreed. Did Carleton Coon do anything in Asia besides hunt monsters? As with any study of intelligence operations, the hidden motives, demagoguery, secret documents, sealed archives, and the fog of the cold war make it difficult to be sure just what happened until more official documents are made available. To add one last intrigue to the story, this author received an interesting reply to attempts to procure Freedom of Information Act requests from the United States government over this issue. The CIA supplied once-classified documents on the intelligence work of Dillon Ripley, Carleton Coon, and others. Whenever I made a search term request they happily gave me what they had or requested further information to help them make a more thorough search, then sent that material along. When I made requests for information they did not have, I received almost apologetic letters stating they did not have any materials on that topic. When I made a request using the search terms, YETI and ABOMINABLE SNOWMAN, however, they did not say they had no such materials. The terse and cryptic reply stated "the mission of the Central Intelligence Agency is primarily concerned with the collection of foreign intelligence matters that affect the national security of the United States. Therefore, we must decline to process your request."[83] This suggests they have documents on these topics, but they are considered national security, so the information will not be released. Ivan Sanderson's ranting about the importance of getting the Yeti before the Soviets may have had some basis after all.[84]

In the end, using the hunt for the Yeti as a cover for espionage may or may not have produced useful intelligence for the American or British governments; it certainly did not produce useful scientific data about hairy bipeds roaming the wastes of central Asia. The extant correspondence shows that whatever their political motivations, the scientists and amateurs involved in the Yeti hunt were sincere in their desire to gather irrefutable scientific evidence proving the creatures existed. From the beginning of anomalous primate studies in the 1950s, amateurs complained that professional scientists did not take them or their work seriously. Bernard Heuvelmans and Ivan Sanderson made this argument one of the central paradigms of the field. However, while members of the scientific world did dismiss the enterprise, a number of them embraced it. Reports of the Yeti—especially after the publication of the Shipton photos in 1951—immediately captured the imagination of anthropologists and primatologists, who began serious research. They eagerly sought out physical evidence and photos to examine and seriously speculated on the possibilities of the creature's existence. They excitedly tested hair and stool samples as well as footprint evidence. Far from being upset that amateurs brought such material to them, they encouraged it. Carleton Coon's pique regarding Tom Slick came not because Slick asked him to go on a Yeti expedition to Nepal, but because at the last minute Slick changed his mind and *did not* bring him along. George Agogino grew upset with Slick because his methods and approaches made it difficult for Agogino to study the Yeti: something Agogino keenly wanted to do. William Charles Osman-Hill eagerly sought the Pangboche Hand for study. Troubled by the behavior of some of the amateurs—and the lack of any followup evidence coming out of Nepal—the academics who wanted to investigate the phenomenon and thought it a worthwhile activity, drifted away and went back to their other work. The frenetic style of the amateurs, the scanty evidence that did not hold up to their microscopes, blood tests, and wide knowledge of animal behavior, anatomy, and evolution, combined to put the scientists off. The Yeti saga highlights the difficulties that arise when scientists and enthusiastic amateurs with less sophisticated training investigate the same phenomenon. The scientists asked for legitimate evidence and applied to it the research techniques of zoology, biology and evolution studies. They urged caution and waved off worthless evidence. Amateurs often interpreted this normal scientific behavior as stonewalling, or disdain for the search, or as evidence of a grand conspiracy to stifle the truth.

Eggheads like Osman-Hill, Coon, Agogino, and their fellows wanted facts, but the crackpots brought them none (even when the *Daily Mail* Expedition did bring a few specialists nothing turned up). What they received for inspection proved frustratingly worthless. As it was, they resisted coming out too publicly on the issue. With all the public accounts of the expeditions to Nepal, the scientists remained relatively obscured: Agogino advocated keeping their presence quiet until final reports could be prepared. Newspaper articles from this period generally refer vaguely to "the scientific community" rather than to any individuals. The most public scientist at this time was the American, Marlon Perkins. He was involved only briefly at the end of the Asia phase of the story and was more a popularizer—and a skeptic—looking for material for his television show rather than as a formal researcher. Some of his antics, like wearing the Yeti scalp as a hat in a widely circulated photograph, proved as annoying and counterproductive as anything Tom Slick did. Agogino did not like the "Hillary Mob" any more than he liked the Slick group.[85] The academics preferred materials collected under carefully controlled conditions and by scientific techniques as opposed to the collection methods Slick and his team engaged in.

With the Hillary-Perkins expedition, the era of big Yeti hunts ended. Many of the serious academic scientists who had rushed with enthusiasm to the Yeti cause had their enthusiasm sorely tested. Most laid aside their interests in hairy monsters, at least publicly. For a group of amateurs in North America, however, the enthusiasm had just begun.

# Chapter 3

# Bigfoot, the Anti-Krantz, and the Iceman

Far from the swashbuckling mountaineers of the Yeti hunts of Asia, in North America the model of the dedicated, obsessive, amateur manlike monster hunter took the form of Swiss-born, Canadian René Dahinden (1930–2001). The irascible and original Dahinden entered the quest for Sasquatch upon his arrival in the New World and developed a reputation for being coarse and abrupt with anyone he thought a fraud or a fool: which in the end formed a rather long list. He regularly discounted scientists, whom he variously referred to as boffins and deadheads. To his credit, with no scientific or research training—in fact little formal education at all—he launched himself into the Sasquatch fray and tirelessly tracked down and documented witnesses and went to the sites of alleged direct encounters. Like many amateurs he secretly yearned to be an academic and live the life of an erudite scholar, but felt hurt, embittered, frustrated, and resentful of those who did: especially when most of them dismissed him. He sacrificed his family life and stability in order to continue his chase and never looked back or showed much public regret. Not as sophisticated as Bernard Heuvelmans, as well published as Ivan Sanderson, or as skilled a writer as John Green, he pursued other monster hunters as tenaciously as he pursued monsters. While eccentric, he was not an out-of-control paranormalist and approached his subject as rationally as any academic should. He represents a great lost opportunity in the history of science. He had the heart, but not the tools. A prime mover in the search for anomalous primates, he could lay claim, along with Green and Peter Byrne, to the title of "grand old man of North American monster hunting.

## The Anti-Krantz

René Dahinden and Grover Krantz represent the archetypes of the crackpot versus egghead concept. Dahinden was born in Lucerne, Switzerland, and deposited by his mother in a local Catholic orphanage. This began a childhood of stepparents, foster homes, and work farms. As a result he developed a penchant for self-reliance and wariness of the motives of others. His one book, *Sasquatch* (1973), written with Don Hunter, is in large part an autobiography. In it he grimly refers to a day at the orphanage when "some people came and picked me up like a dog from a kennel."[1] On his own by the early 1950s, René happily roamed about postwar Europe. In 1953 he immigrated to Canada with his future wife, Wanja Twan, and found work on a dairy farm in Calgary, British Columbia. Shortly after his arrival Dahinden read a newspaper article about the expedition sponsored by the *Daily Mail* to the Himalaya Mountains in search of the Yeti.[2]

In *Sasquatch*, Dahinden explains how the idea of the *Daily Mail* Expedition fired his imagination. He told his employer, Mr. Willick, that he would love to go on such a trip. Willick offhandedly asked why he would want to go all the way around the world to look for something wandering around right in his own backyard. Long tradition said a creature just like the Yeti lived in Canada: a hairy thing or some such that the native people believed in and which non-Indian Canadians occasionally encountered as well. Dahinden began searching local libraries for information on the creature. By 1956 he had made the acquaintance of John Green, a local newspaperman, also interested in the Sasquatch legend. Green, a university trained journalist, published the *Aggisiz-Harrison Advance* and lived in Harrison Hot Springs, British Columbia. While interested in the legend himself, Green told Dahinden not to take it too seriously. In 1957 the town fathers of Harrison Hot Springs hatched a scheme to cash in on the Sasquatch legend now somewhat more popular because of the *Daily Mail* Expedition. They wanted an expedition of their own— mostly as a publicity stunt—and both Dahinden and Green took part. Though the foray never came off, Dahinden and Green established themselves as local experts on anomalous primates, their private interest now a public career. Unfortunately, the circus atmosphere and the characters drawn to the production annoyed the serious Dahinden, and his contempt for fellow Sasquatch hunters began. In 1959, as Carleton Coon, George Agogino, and other scientists' interests waned, Dahinden's career found a focal point when large, human-like Sasquatch footprints were found near Bluff Creek, California.

Dahinden and Green headed there, eventually encountered Tom Slick, and became caught up in Slick's latest adventure, the Pacific Northwest Expedition.

René Dahinden had always been a solitary man, keeping his own council and going his own way, happiest when leading his own life. The Harrison Hot Springs and later operations showed he did not mix well with others, especially concerning his favorite topic. It would not be long before he jettisoned everything in his life—including his wife and children—which did not add to, or benefit, his Sasquatch work. He spent the early 1960s working out a situation to allow him three things crucial to his Sasquatch hunt: a minimum level of financial security, a life unencumbered by anything holding him back, and plenty of free time to pursue his work. He and his wife had been running a boat rental business, but as he did not deal well with tourists it faltered. He then hit upon the idea of recovering used and lost golf balls from driving ranges and golf courses and buckshot from shooting ranges for resale. It was ideal for Dahinden's personality: He could work alone, earn money, and set his own hours. He eventually worked out a deal with the Vancouver Gun Club so that he could recover shot from its ranges and permanently park his mobile home on the club's grounds. (He lived there in his trailer the rest of his days). Though he had put his financial life in order, his personal life fell apart.[3]

His long trips away from home to interview people and look at sites where the creature had been seen, coupled with his personal nature, took a toll on his wife. In 1967 he told Wanja of his plans to go back to the Sasquatch hunt after a period away from it. Though she had initially been eager to participant in René's hobby, Wanja responded that he had to choose between her and their two sons and his obsession with the bogeyman. They separated and then divorced. In *Sasquatch*, Dahinden recounted with brutal honesty what broke up his marriage. As an indication of his devotion to Sasquatch he dismissed his family once and for all by saying the breakup had been a good "omen," and that it had come at an opportune time. No wife or children to care for made his search for the elusive "big hairy bastard," as he called it, that much easier. He pursued his quest with a level of intensity and single-mindedness to make Captain Ahab proud. He spent extended periods exploring the wild and beautiful forested regions of western Canada in solitary contemplations and bucolic musings. Whether he found the creature or not the search kept him alive and nurtured his soul. Alone in the hills with nothing more than a backpack and a rifle, or alone in his trailer with his classical music, pipe, and books, he inhabited a one man universe, and few were invited in for very long. In this way he could live the fantasy of rough, independent mountain man and independent scholar. Reading his own descriptions of his activities, one gets the feeling he preferred this lifestyle over any other, with no one bothering him, holding him back, or wasting his time. Humans always disappointed him, let him down; nature never did with one glaring exception.

## Enter Bigfoot

In a trajectory that mirrored the story of the Yeti, the legend of Sasquatch began as a local one. Few outside the Pacific Northwest or the small community of manlike monster enthusiasts knew much of the story. Most of the country focused on things like the possibility of squadrons of Soviet bombers appearing over their cities or the air-raid warning—that annoying long beep of the Emergency Broadcasting System—suddenly coming on their radio or television at a time other than noon on Saturday. The postwar years of the 1950s proved fertile ground for the germination of belief in unusual phenomena. It is during this period that interest in UFOs blossomed. Following Kenneth Arnold's much publicized 1947 encounter with peculiar flying machines—soon dubbed flying saucers—around Mount Rainier, Washington, a growing body of people, especially in the western United States, claimed to have seen and encountered close up, contraptions and their occupants from other worlds. The UFO phenomena began its modern life in the skies over the forests where the Sasquatch lurked below (later monster enthusiasts would see a direct connection between flying saucers and Sasquatch).[4] The eyes of most followers of the outré turned to the skies over the Pacific Northwest, not down into its forests. The word Sasquatch is of Native American origins. First brought to prominence in the 1930s by Chehalis Washington Indian Reserve agent and teacher J. W. Burns, it is thought to be an anglicized variation of a Halkomelem Indian word for hairy giant.[5] Sasquatch, like the Yeti, would have remained a local story if not for the invention of a startling new moniker.

On August 27, 1958, a road construction gang working in a remote part of the thick northern California wilderness near a place called Bluff Creek, came across something unusual. During the night someone, or something, had left enormous footprints in the dirt near their bulldozer. The juxtaposition of the modern machine and the seemingly ancient, but recently made footprints appeared otherworldly. The men found more tracks over the next few weeks, though they never saw what made them. One of the workmen fascinated by the tracks, Jerry Crew, contacted a local taxidermist, Bob Titmus, and asked for instructions on how to make plaster casts of the prints. Armed with instructions, Crew soon had a giant plaster foot. He took it to local journalist Andrew Genzoli of the *Humboldt Times*. Genzoli had heard legends of wild men roaming the forest. When Crew told Genzoli the men on the construction gang had taken to calling the unseen visitor *Bigfoot*, Genzoli wrote up the story and put it on the Associated Press wire service. The swift public reaction caught them all off guard. As Genzoli put it, "it was like loosening a single stone in an avalanche."[6] The name carried a media punch to make any advertising executive green with envy. An insouciant turn of phrase by construction workers had given birth to a phenomenon. Suddenly, Bigfoot stories started coming out of the forest woodwork. Such creatures had been seen for decades, even centuries, in North America, but now a new generation undertook an investigation. That generation included Canadian journalist John Green, Swiss immigrant René Dahinden, and oil man turned monster hunter, Tom Slick.

Hearing what had happened in California, Ivan Sanderson wrote up a version of the Bigfoot story for *True* magazine and sent it to Slick. Disappointed by his Yeti adventure, Slick determined to catch its American cousin. John Green hurried to the site, hoping to get a sense of the trail while it was still warm. Green had yet to embrace the idea of Sasquatch completely. Seeing the tracks at Bluff Creek, however, left him with no doubts the creature existed. He talked to Jerry Crew and Bob Titmus, who had by then found more tracks. Plaster casts Titmus made found their way to Tom Slick. Excited by the evidence, Slick rounded up some of his former cronies. The momentum pulled in Green, Dahinden, Titmus, Sanderson, and even George Agogino, who had Bob Titmus send plaster Bigfoot casts to Carleton Coon for inspection. Slick's blood quickened, and so the Pacific Northwest Expedition began. Later, Sasquatch enthusiast Ivan Marx (of future Bossburg infamy) and Ray Wallace (the foreman of Jerry Crew's construction gang) joined the group as well.[7]

Tom Slick also contacted his old Yeti hunting friend, Peter Byrne, and asked him to come along. Byrne first met Green and Dahinden through Slick. Over the coming decades these three men would step to the fore as the most prominent and respected researchers involved in anomalous primate studies. They could not have been more different: Green, level headed, rational, a sober thinker and writer; Byrne the friendly talkative Irish adventurer always ready with a big smile and a great story, full of optimism; and Dahinden the taciturn, secretive

Swiss bulldog. They all took a serious and individual approach, having their own ideas and plans. Their relationships would go through periods of turmoil as they danced around each other and the question of how to seek such a creature and what to do with it if they found it. His experience with the Harrison Hot Springs project, led Dahinden to scorn the expedition approach to Sasquatch research. Despite these misgivings, but with a need for an income, he took a paid position with Slick's enterprise. The Pacific Northwest Expedition began with great enthusiasm but deteriorated quickly into name calling and internal squabbles. The North Americans resented the outsider Byrne being brought in to lead them; others thought Dahinden concealed information from them; some had ulterior motives for getting involved, and most had no idea how to go about capturing a monster. Dahinden, disgusted with it all, considered his fellows either stupid or idiots, and let them know it. Some of the team members' antics, like hanging meat and fruit in trees to attract the animal, drove Dahinden crazy. He left the group after a short time and, disillusioned with the Sasquatch-hunter community, left the field briefly to focus on other things. John Green kept at it, now thoroughly convinced Sasquatch was real and, after a decade of research, published the influential *On the Track of Sasquatch* (1968). Like other monster missions the Pacific Northwest Expedition turned up nothing.[8]

Tom Slick may have been willing to go out on a limb to find monsters, but he also had a knack for alienating others he brought up in the tree with him. Like Carleton Coon and George Agogino before him, Ivan Sanderson—whom Slick had asked to become part of the Pacific Northwest Expedition—was increasingly annoyed by Slick's behavior. Sanderson wanted to go, but his schedule prevented it, as did his financial situation. Fearing becoming indebted, he refused to allow Slick to cover his expenses. Sanderson wanted to show up when he wanted and follow the expedition wherever it might be. Slick had no problem with that, but had some special restrictions. Sanderson could come and go as he wished and "go ahead with the normal practice of [his] profession," but 90 percent of any money Sanderson received for his writing had to go into the expedition's coffers.[9] As Sanderson's only source of income was his writing, he refused. He wanted to be able to write whatever he wanted for whomever at any time and keep his earnings. "I could not agree to any restrictions of any kind in my writings," Sanderson told Jeri Walsh, a Slick confidant.[10] Sanderson worried about his reputation as well. After looking into the Bigfoot incident of Jerry Crew in 1959 he submitted an article to *True* magazine. The magazine made editorial additions to the article without Sanderson's knowledge, which he felt distorted the point he had worked so hard to make. He angrily told journalist Andrew Genzoli that he was "utterly disgusted" by the "distortions and unauthorized insertions that I did not write" that the *True* editor put in.[11] Four years later, folklorist Lynwood Carranco, unaware of what Sanderson claimed, referred to his article as being "told in a sensational manner with many inaccuracies."[12] Bad enough to be labeled a crank for your own work, but to suffer because of an editor's tinkering galled Sanderson, who took great pride in his journalistic abilities.

In addition to the writing gag order, Sanderson worried over Slick's field techniques. Having been on and led numerous scientific research expeditions in the past, Sanderson was troubled. Just because Slick fronted the money for the expedition did not mean he should lead it. Scientific field operations had accepted rules for how to proceed, rules designed to make them run smoother and accomplish more. Having the money man out in front, in Sanderson's view, hampered how the team worked and adversely affected decision making. "All his [Slick's] troubles," Sanderson said, "stem from his pure lack of knowledge" of the rules of scientific expeditions. "He'll never get anything but headaches and wastage of money if he doesn't adopt these methods."[13] Sanderson and Slick tried to work out a compromise. They came to an agreement that said Sanderson would not write about the California expedition as long as Slick or his team was engaged in field work. He could go into the field himself and write about Sasquatch, but just not mention Slick's project. Sanderson would also supply written guidelines for team members to aid in the search for the Sasquatch: what Sanderson referred to as the "Bukwa." Not wanting to have lawyers become involved in their deal, Sanderson asked Slick to simply sign a copy of his letter of understanding if he agreed with its contents. Slick signed and returned the unofficial contract.[14]

Sanderson had other concerns. Expedition member Ray Wallace caused problems with accusations about Sanderson that enraged the Scot. He may have accused Sanderson (who was not sure what Wallace's "accusations" actually entailed, having heard them only third hand) of not living up to his deal with Slick and publishing materials on the Slick expedition. That Slick began to distance himself from Sanderson because of the allegations caused Sanderson even more annoyance. In a memo for his own files, Sanderson said, "I have adhered to my part of the agreement [with Slick] faithfully but I regret to say Mr. Slick has not in so far as he has not kept me informed of the activities and findings of the expedition" the way Slick had agreed he would. Sanderson said Wallace's statement "was a malicious lie" and threatened a lawsuit.[15] He then wrote directly to Slick to say that "this man Wallace is a menace and always has been."[16] Wallace went on to have his fair share of Sasquatch related troubles. He admitted faking tracks, and shortly after his death in 2002, his family announced that he had faked the Bigfoot tracks found by Jerry Crew: an accusation vehemently opposed by most of the anomalous primate community, who claimed personal familiarity with the various characters involved.

Sanderson would not have to worry about whatever deals he made with Tom Slick for much longer. It all came to an end in 1962, with Slick's sudden death in a plane crash in Montana. George Agogino told Carleton Coon that "our old friend Tom Slick died this weekend in a Montana plane crash." Agogino, who began as a skeptic but came to believe anomalous primates were real, felt with Slick a huge opportunity had been missed. He lamented to Coon that "most of his [Slick] money was spent on pure trash research."[17] Ivan Sanderson disillusioned Agogino as well. Agogino wrote the introduction to *Abominable Snowmen* without first reading the manuscript and later regretted it, feeling the book would only inspire popular writers not academics. Agogino argued that serious research into anomalous primates needed to be done, but would only get done if the amateurs toned down their sensational and unscientific behavior. To him, the amateurs had simply gotten in the way and derailed the entire effort. He questioned Sanderson's habit of labeling every historic reference to wild men or anything similar as descriptions of anomalous primates. Wearily, he sighed, "rather soon Sanderson will claim Ishi was really a Yeti"[18] (His reference being to the lone surviving member of a vanished Native American Yahi tribe discovered in California in 1911).

With his death a fog fell over Tom Slick's work. The various expedition teams broke up, and the materials they collected went to the Slick estate. From there, only legends and dark whispers remain of what happened to all the materials collected in his name and paid for with his money. Researchers have had a difficult time getting at his papers. Slick's biographer, Loren Coleman, managed to ferret out a good bit of material, but could take the search only so far. In *Tom Slick and the Search for the Yeti* (1989), and the later edition, *Tom Slick: True Life Encounters in Cryptozoology* (2002), Coleman went to considerable lengths to make Slick come to life, but the Texas oilman with a preoccupation with monsters and possible CIA connections remains a cipher who continues to fight against the pick-locks of biographers.

## John Napier

One of the most thorough academic investigations into the topic
of manlike monsters to see print at this early stage came from then
Smithsonian Institution primatologist John Napier (1917–87).
Having studied medicine in the United Kingdom, Napier became
especially interested in human foot and hand anatomy. This focus
resulted in his invitation, in the early 1960s (as the great Yeti expeditions wound down), to make a study of the forelimbs of the recently
discovered fossil *Proconsul*. The monograph he produced on this
material established Napier as an important researcher into primate
origins and locomotion. In 1964 he assisted Philip Tobias and Louis
Leakey in the description and naming of the African hominid *Homo
habilis*. Prestigious appointments in the United States and England
followed.[19] A man of wide ranging interests (he loved to perform
as a magician), Napier followed anomalous primates literature with
interest. While at the Smithsonian Institution's National Museum of
Natural History he took a leading role in the case of the Minnesota
Iceman and screened the Patterson-Gimlin film of Bigfoot. He regularly examined photographs of footprints and casts. When he saw the
Cripplefoot prints in photos he, like Grover Krantz, felt they were too
good to have been faked. He considered Bigfoot "the most interesting
of all the hairy men" because, he said "it is cast in a human image."
He told Roger Patterson that postulating Bigfoot crossing the Bering
Strait land bridge made sense. Noting the common wisdom of the
day, that humans had entered the Americas only ten thousand years
ago or so, he mused that "it had always been a matter of surprise to
me that he waited so long!"[20] By the early 1970s, Napier had reached
a point where he started working on a project of his own.

When he began writing his book, Napier contacted fellow British
primatologist William Charles Osman-Hill, who suggested he contact
the estate of Tom Slick. Napier wanted to see a "confidential report"
the Slick camp had produced, but not published, which summarized
much of the work done in Nepal, and which Osman-Hill, but few others, knew of. Napier's initial contact with the Slick estate went well.
He assured the trustees of the Tom Slick Foundation that as a serious
researcher "my feet are—by upbringing—firmly on the ground."[21]
The return letter from an "Independent Executor of the Estate of
Tom Slick" cheerily responded that materials would arrive shortly.[22]
When they did, the package did not include the report Osman-Hill
mentioned. A follow-up letter specifically asked for the confidential report. This time a less cheery reply said, "We regret we cannot

furnish the information requested." The uncollated and uncataloged nature of the material prevented it.[23] Napier said he understood and asked for "a further favour" of some specific photographs from the expeditions. The answer to this request stated curtly that he could not have this material "due to [the] possible interest of other people, some known and some unknown, in Mr. Slick's expedition."[24] Confused by the reply, and growing exasperated, Napier wrote back saying,

> The late Mr. Tom Slick, like so many other enthusiasts of this phenomena, often expressed his disappointment that scientists in official standing did not bother to interest themselves in the Abominable Snowman or Bigfoot affair. Yet when I, as a reputable, internationally known scientist in the field of primate and human evolution, request simple, *factual* information you deny it to me.[25] [Napier's accentuation]

No further Slick Foundation replies arrived, and he never received the information he requested. Napier later said, "It has proved difficult to find out exactly what the Slick Expedition *did* [his italic] find." He further stated that Slick's "executors are remarkably taciturn."[26] Despite running into a dead end with the Slick Foundation, Napier went on to write *Bigfoot: The Yeti and Sasquatch in Myth and Reality* (1973). He looked at the Sasquatch evidence with the eye of an anatomist. He felt that while some of the evidence intrigued him, he had not come to a firm conclusion. He seems to go back and forth in his assessment. Examining the foot and hand anatomy, Napier deconstructed the idea that Bigfoot or the Yeti represented some kind of relic holdover from the mists of time (something that did not endear him to Russian monster hunters). Looking at *Gigantopithecus* fossils, he said that while the creature would have been large he doubted it a nine-foot giant. He added that no real evidence proved *Gigantopithecus* a biped either. Even if it looked exactly like a Sasquatch, the lack of recent *Gigantopithecus* fossils suggested against its extending into the modern age. Another stretch of logic concerned Sanderson's pet theory that *Pithecanthropus* and Neanderthals represented the source of the Yeti. Both creatures had progressed in a generally human direction, Sanderson said. They had mastered fire, had early forms of spoken communication, and nibbled at the edges of tool making and culture. For populations of these creatures to suddenly turn from that course of development to grow into hairy, solitary giants without any trace of cultural advance made no sense from an evolutionary point of view. In other words, *Homo erectus*, and certainly the Neanderthals, had advanced too far along to have devolved into Bigfoot. Though he felt none of the popular theories for the origins of Bigfoot were tenable, he did admit the one theory that had any kind of chance was that of the *Gigantopithecus*.

In the end, however, Napier put forward as much evidence *against* these creatures existing as he did *for* them existing. Just fobbing Sasquatch off as a Neanderthal or feral human, or even as *Gigantopithecus*, only counted as an easy way out and did not fit the facts or simple logic. "By postulating that a monster is a relic form—a hangover from the past—monster fans," he argued, "feel absolved from the necessity of explaining how such an outrageously unsuitable creature has evolved in the light of present-day ecology."[27] Resolving the problem once and for all (and coaxing the scientific community back into the fray) demanded two things: a theory to explain the creature's existence, and a body—or at least part of a body—to confirm the creature's existence. The most popular explanatory theory involved a creature long extinct and would bring anomalous primate studies into the realm of competing human evolution arguments. For a brief moment, a body arrived on the scene as well.

## The Gigantopithecus Theory

Two men leaned silently over a glass topped, refrigerator sized box in the darkened interior of a trailer parked inside an old, rural barn. Stunned by what they saw, neither man could speak. Though the Minnesota winter had yet to officially begin, snow already coated the ground, and the men made little clouds of steam as they breathed. Staring intently at the thing the box contained, Ivan Sanderson and Bernard Heuvelmans gazed at each other silently, as if to ask, "Can this really be what it looks like?" Inside the box was a block of ice containing the carcass of what appeared to be an ape-man.

Still excited months later, Sanderson would write that he saw "a comparatively fresh corpse, preserved in ice, of a specimen of at least one kind of ultra-primitive, fully-haired man-thing, that displays so many heretofore unexpected and non-human characters as to warrant our dubbing it a 'missing link.' "[28] Despite the cold and their excitement at encountering whatever this was, Sanderson and Heuvelmans worked at keeping their composure and maintaining an air of scientific objectivity. They carefully kept notes, made drawings, and took photographs. The clicking of Heuvelmans's camera made the only noise in the confined space of the trailer. After years of reading stories of other people's encounters and sightings, examining drawings and photos, the two men finally had the body of an anomalous primate, an ABSM, a manlike monster, all to themselves just a few inches away under glass and ice.

And therein lay the rub. Even if the trailer's owner, a carnival exhibitor named Frank Hansen, had wanted to take the glass top off—and he didn't—several inches of ice wrapped the creature like a crystal shroud, preventing an intimate examination. Clear in some spots, cloudy in others, what could be seen of the face had a sad, strained, closed-eye expression, as if it had expired under painful circumstances. The head seemed to have a wound, the left arm raised in what looked like a feeble attempt to ward off an attacker. In his block of ice, in a quiet, cold, rundown Midwestern barn, far from the image of a frightening monster, the thing possessed an almost touching melancholy.

If real, the Minnesota Iceman, as the frozen artifact was now being called, filled in only half the equation. Even with a body as proof, a full explanation required a convincing evolutionary theory to explain it. In their pursuit of the Sasquatch, the amateur naturalists went out into the field to focus on verifying the animal by tracking its movements, casting its tracks, and photographing it if possible. At the same time, others speculated on what might account for the existence of such a creature. If large bipedal apes cavorted about the big timbers, they had to have come from somewhere. The British naturalists who commented publicly on the Yeti in the 1950s asked such questions about the creature's origins. J. P. Mills, who consulted on the *Daily Mail* Expedition, mentioned "that Central Asia did in prehistoric times contain a primitive species of man is proved by abundant finds of remains of the so-called Pekin [sic] Man" and so the idea of a Snowman fell within the realm of possibility.[29] As for Sasquatch, authors Don Oakley and John Lane looked to Java Man as an example of what Sasquatch might be related to, and brought up *Gigantopithecus*. They saw the big ancient ape as proof that "giants really did exist in the past."[30]

That giants may have roamed the New World in years gone by found support through the existence of the Bering Land Bridge connecting Asia to North America. An early theory to explain where Sasquatch came from coalesced around the idea that, whatever its antecedents, it came to North America from Asia by walking. The general consensus of mainstream paleoanthropologists, however, held that the first hominids to cross into the New World did so as fully developed *Homo sapiens*, not evolutionary throwbacks or monsters. Human remains have been found in the Americas that discoverers claimed to be fossils, but Lansing Man in California, Vero Man in Florida,

Trenton Man in New Jersey, and Nebraska Man all turned out to be misidentifications of modern skeletons. In Argentina anthropologist Florentino Ameghino (1854–1911) discovered what he thought to be ancient humans, whom he called *Tetraprothomo* and *Diprothomo*. A number of claims of the discovery of *Homo erectus* fossils in the Americas have all gone unverified.[31]

One theory that the first hominids to enter the Americas might have been something other than modern humans came, in 1947, from the amateur anthropologist and adventurer Harold Gladwin (1883–1983). After striking it rich on Wall Street, Gladwin abandoned that life to pursue his passion for New World archaeology. His enthusiasm and money allowed him to do some genuinely useful work on ancient American peoples, particularly in the area of chronological data based on pottery design and tree ring analysis.[32] Musings on the origins of the Indians in *Men Out of Asia* represented his less respectable work. The book had an introduction by Carleton Coon's teacher, the noted Harvard anthropologist Ernest Hooton. A eugenicist who wrote books with titles like *Apes, Men, and Morons* (1936), Hooton did work in population planning and, during the rise of the Nazis, suggested that in order to clean up American society something had to be done to stop the breeding of "protected inferiors."[33] Though he lauded Gladwin's book, Hooton admitted that his ideas about protohumans in America were ones that "I myself do not agree with."[34] Gladwin argued, among other things, that the idea that all men are biological brothers "has some rather obvious flaws."[35] His descriptions of various human ethnic groups are racist and paternalistic, his pithy prose insulting (which may account for Hooton's admiration). Gladwin says that Black people can trace their lineages back to the cavemen, but Whites are "amply justified in refusing to recognize" Java Man or Peking Man as ancestors. He then wonders why the Neanderthals were removed from the human family tree when at the same time it includes "the modern low-brows in our sapiens family."[36] The racist garbage aside, Gladwin postulated, based on no evidence, that late surviving Neanderthals, and even *Sinanthropus* and *Pithecanthropus* populations, entered the Americas and were ancestors of the modern aboriginal people.[37] While he did not discuss Sasquatch, his speculations about "relic" populations are consistent with later Bigfoot origins theories involving non–*Homo sapiens* hominids entering the Americas.

The most popular and lucid theory explaining the Yeti or Sasquatch, and the one Grover Krantz embraced and worked to spread, related it to *Gigantopithecus*. In the 1930s German anthropologist Ralph von

Koenigswald (1902–82) discovered the fossil remains of this creature in China. He had taken to prowling apothecary shops, looking for fossils. Pharmacists and herbalists made popular folk remedies by grinding up fossils, called "Dragon Bones," for use as various medicines. On one of his trips he found several large primate-like teeth of a kind he had never seen before. Because of their primate morphology and large size he called them *Gigantopithecus*. At the outbreak of World War II von Koenigswald handed his fossil collection off to colleague and fellow China fossil expert, Franz Weidenreich (1873–1948), in order to get them out of China before the marauding Japanese army could seize them. An attempt by the same group of European and Chinese scientists to get the remains of the first Peking Man fossils out of China did not fare as well, and the remains disappeared into legendary oblivion.[38]

While Ralph von Koenigswald speculated on the nature of *Gigantopithecus* Weidenreich made the most extensive description of it. Weidenreich believed all hominid fossil material, including that of both the direct ancestors to humans and modern humans, themselves represented one long line of descent, or continuity. He said, "Not only the living forms of mankind but also the past forms—at least those whose remains have been discovered—must be included in the same species."[39] As part of this continuity, *Sinanthropus* (Peking Man) and *Pithecanthropus* (Java Man) must both be considered "a true man." Building upon the work done in Palestine by Grover Krantz's future teacher, Theodore D. McCown (1908–69), Weidenreich argued that the Mount Carmel, Galilee, skull found at Skhŭl cave by McCown in the 1930s represented an intermediate stage between the Neanderthals and archaic modern humans.[40]

Weidenreich concluded that the big teeth had come "from a giant man and should, therefore, have been named *Gigantanthropus* and not *Gigantopithecus*," in order to show their closer relationship to humans.[41] The *Gigantopithecus* teeth displayed morphological consistencies with those of *Sinanthropus* and *Pithecanthropus*.[42] Therefore, he argued, *Gigantopithecus* belonged in the human line. Furthermore, he asserted the fossils from Asia "prove that there was never a single cradle of mankind—whether located in Asia or Africa," but that humans evolved independently and simultaneously in different parts of the world.[43] Making a "very general statement" about the overall appearance of *Gigantopithecus,* he said the big teeth suggested a big jaw, which suggested a big skull that would require a rather large body to support it. He reasoned that since *Sinanthropus* and *Pithecanthropus* were bipeds, *Gigantopithecus* probably stood as a biped as well. Weidenreich concluded that it would have had an overall massive humanlike body plan: in other words, a big hairy humanoid.

Not long after Weidenreich and von Koenigswald published their work on *Gigantopithecus*, it found a life as a way to explain the Abominable Snowman. While Carleton Coon and George Agogino privately discussed *Gigantopithecus* as the Yeti forebear, Bernard Heuvelmans first applied knowledge of the creature to support a possible origins theory for the Yeti in print in a 1952 article in French.[44] Heuvelmans also made reference to the work of von Koenigswald and Robert Broom (1866–1951), who had found *Australopithecus*-like fossils in South Africa at Swartkrans, called *Paranthropus*. The English translation of Heuvelmans's work, *On the Track of Unknown Animals*, made the idea accessible to a wider English speaking audience, especially in North America, where reports and legends of a creature just as Weidenreich had described had persisted for years. Heuvelmans claimed that "I had already related the legends of Himalayan ogres to von Koenigswald's discoveries," before the appearance of the Shipton photos.[45]

Carleton Coon read Weidenreich's and von Koenigswald's work on the giant ape and made a connection between *Gigantopithecus* and the Yeti. In addition to the Agogino correspondence, Coon's letters to ornithologist Dillon Ripley expressed this opinion the next year (1953). Coon's first public reference to *Gigantopithecus* came in a brief mention in his book *The Story of Man* of 1954.[46] It seems logical that the few academics—like Heuvelmans, Coon, and Agogino—with knowledge of both the latest discoveries and research on human evolution and the Yeti would link the Snowman and *Gigantopithecus*.

The anatomist Vladimir Tchernezky, of Queen Mary College, London, studied the Shipton photos independently and also put forward *Gigantopithecus* for consideration. Tchernezky felt for sure an unknown primate wandered around Nepal. He said excitedly that the photos showed "a new form of higher anthropoid will soon be discovered." He went on that "this form will be related to the fossil Gigantopithecus."[47] Several years later Tchernezky did a detailed study of the Shipton pictures by recreating them in three dimensions for an article in *Nature*.[48] Heuvelmans thought it not impossible for a *Gigantopithecus*, or *Meganthropus*, or *Paranthropus* population to have moved to the high country of Tibet and survived. Heuvelmans, who felt the simultaneous development of *Gigantopithecus* theories by Tschernezky and himself (he seems to have been unaware initially of Coon's mention) gave greater weight to the idea. He suggested the Yeti be given the scientific name *Dinanthropoides nivalis*.[49]

While not ready to make a definitive statement on the subject, Heuvelmans thought he had something. He said the Yeti/ *Gigantopithecus* link "is utterly hypothetical, [but] provides the only entirely acceptable explanation of the mystery of the Abominable Snowman." It also fell in line with colleague Ivan Sanderson's contention of multiple anomalous primate species existing in various parts of the world.[50] While Heuvelmans may have been the first to establish in print the paradigm of the Yeti being a *Gigantopithecus* relative, he did not pursue the issue beyond stating it. Few researchers did.

The *Gigantopithecus* theory would remain unproven until some physical evidence arrived to support it. The competing theory of hominid relic, where small populations of Neanderthals somehow managed to survive to the modern era, had its supporters as well. Ivan Sanderson and Bernard Heuvelmans both felt the many different descriptions of anomalous primates from around the world suggested Snowman diversity. In this way no clash broke out between *Gigantopithecus* and relic theory. The concepts could be held simultaneously and without contradiction. The relic theory appealed to Russian investigators like Boris Porshnev. In 1968 and 1969 the relic theory seemed that much more plausible because a relic appeared.

### The Iceman Cometh

The body of a Snowman or Sasquatch would end the question convincingly and quiet all those annoying skeptics who called anomalous primate enthusiasts crackpots and condescendingly patted their heads like befuddled children. Monster hunters looked in Asian mountain ranges and the forests of the Pacific Northwest. Then a body turned up, not in a way anyone would have anticipated, yet in a way that seemed so appropriate—in traveling carnival sideshow.

From the 1940s through the 1960s Ivan Sanderson had written a series of articles on the Sasquatch and Yeti mysteries for some of the leading men's magazines of the day like *Argosy* and *True*, which influenced and inspired the early mystery-ape researchers. His 1946 article inspired Bernard Heuvelmans and his 1959 article prompted Roger Patterson to grab a camera and head to California. He mentioned *Gigantopithecus* in his *True* magazine article "Abominable Snowmen are Here!" of 1961 as circumstantial evidence for Sasquatch. So it was not unusual when a young aspiring scientist called him to report something strange found at a local fair.

The Minnesota Iceman floats through the lore of anomalous primate circles like a ghost and has been a source of constant debate and conjecture. The changing story of the central character—retired air force officer Frank Hansen—cast doubt on the entire enterprise. The spuriousness and eventual "disappearance" of the Iceman itself usually makes the story only a footnote, but it is useful as an example of how mainstream scientists did take such creatures seriously and why, afterward, they tended to avoid them. Unlike most monster hoaxes, this one did not start out as something meant to fool researchers the way the infamous Piltdown Man had intentionally fooled British naturalists in the early twentieth century, or the 2008 Georgia Bigfoot hoax intended to fool believers. The Iceman started off as a sideshow attraction to entertain and titillate carnival spectators, not Sasquatch enthusiasts. Hansen had read the literature on ABSMs but likely used it to make a convincing gag rather than to fool experts. The serious, but unintended, interest by scientists in the Iceman contributed to its owner suddenly and skittishly making it disappear. Researchers took the wheeze far more seriously than they should have, and it got out of control. For Hansen, what began as a publicity stunt turned into a possible criminal charge that scared him as much as his creature scared rubes at the carnival.

In 1968 Ivan Sanderson lived on a small farm in northwest New Jersey and, in addition to his other cryptozoological activities, ran the Society for the Investigation of the Unexplained, known to its members as SITU). That December the society was contacted by Terry Cullen, an aspiring young naturalist from Milwaukee who said he had seen a hairy hominid-like creature in a block of ice at the Chicago International Livestock Exposition.[51] This roused Sanderson to go have a look. By coincidence, Bernard Heuvelmans had a few weeks earlier arrived at Sanderson's farm for a visit. Heuvelmans had been heading to South America and stopped off in New York along the way. His *In the Wake of the Sea Serpents* had just been released that October in English translation in America by the publishers Hill and Wang. Arthur Wang and his wife held a reception for Heuvelmans in their Riverside Drive, Manhattan apartment to celebrate on October 15. Local publishing and science luminaries, including Willy Ley, attended.[52] Sanderson invited the Frenchman to spend an extended period with him and threw Heuvelmans an informal party of his own, inviting among others, his friend Dr. Hobart Van Deusen from the American Museum of Natural History.[53] Heuvelmans happily accepted the offer to spend a week or two relaxing in New Jersey with his old friend before heading off again. He had no idea his reverie would be interrupted by an unscheduled trip deep into the American heartland to look at a frozen ape-man.

With Terry Cullen's alert, these two fathers of cryptozoology headed to Minnesota. Hansen at first expressed reluctance to show the two investigators his treasure, but finally relented. Hansen told Sanderson and Heuvelmans he had received the creature from a "millionaire" who acquired it in Asia and wanted to exhibit it but keep his identity secret. The two men did their best to examine the ice encased body, sketching and taking photos. The creature looked just as they expected a Neanderthal survival to look (though it appeared to have a bullet wound to its head). What is important is that both men felt the creature genuine and immediately began to spread word of its existence. The Iceman closely fit the theory of a Neanderthal relic. It looked like a stereotypical hairy wild man. More human sized and proportional than the bigger Sasquatch, it fell in line with traditional descriptions of the Almasti of Asia, the Iceman's purported home. Heuvelmans would later opine that it could be related to the Ainu people of Japan. It also validated the multiple ABSM concept and reinforced the notion that the bigger Yeti and Sasquatch originated with *Gigantopithecus* not Neanderthal. Following their inspection Heuvelmans and Sanderson went off to write up and publish their reports.[54]

Ever the hustling author, Sanderson whipped up some extensive notes on what he referred to as the "Hansen Case" and sent them to his old monster hunting cronies, asking them for their opinions. Carleton Coon, George Agogino, William Charles Osman-Hill, and John Napier all provided comments he would later use for an *Argosy* magazine article.[55] Coon received his copy first, as Sanderson personally brought it to his Massachusetts home. Hoping to bring the full weight of the U.S. government to bear on the Iceman, Sanderson also sent copies of his memo to the Bureau of Customs, Department of Agriculture, Department of the Interior, Health, Education and Welfare, and the FBI.[56] He immediately began scheming to get the carcass for himself or for some reputable institution so he could study and also parade it.

The Hansen Case memo is a curious document. At 15 pages in length it seems to have been written with Sanderson in an emotional state. It staggers back and forth between scholarly calm and almost out-of-control agitation over a wide range of issues. After describing—as the later *Argosy* article would—the discovery of the Iceman and telling all he knew of its origins, Sanderson went on at length about the legalities of ownership and customs considerations. He finished with a focused continuation of his career-long rant against the "established and working scientists" who disagree with him. He argued contradictorily that knowledge of the reality of the Iceman could upset the scientific status quo and trigger seismic shifts in modern society. For scientists "this could ruin their reputations if fully proved" and create an "open invitation to all students to ask embarrassing questions."[57] Sanderson believed the scientists were afraid of new ideas and evidence. In his mind, knowledge of the existence of anomalous primates had global repercussions and would undermine the entire scientific enterprise. This fear forced "the Establishment" to ignore, ridicule, or suppress knowledge of ABSMs. Sanderson argued the Iceman would "in any case necessitate rewriting all the textbooks, finally approving Darwin's as opposed to the Christian Church's, theory of our origins."[58] He did not explain why evolutionary biologists would suppress knowledge of something that supported evolutionary biology, rather than church dogma. Not for a minute did Sanderson ever consider the possibility that the reason scientists did not flock to his side was because he had not presented them with convincing enough evidence. The problem always originated on their side of the aisle, not his.

Sanderson's frustration at not being part of the scientific mainstream community came out, again, in one last blast in his Hansen Case Memo. He claimed there was deliberate suppression of the Iceman going on, then concluded,

> We are thus presented with the deplorable picture of the public waiting for just this concrete proof of all that has been drilled into them by both religion and science, while the communications industry drivels, and shakes in its intellectual boots; and the scientists deliberately try to "sink the ship."[59]

In his typically vague style, he never says just which scientists were trying to "sink the ship." In a final irony, after raging against the perfidy of scientists, he did not send the Hansen Case Memo to any of the many amateur monster hunters he knew. He sent it only to scientists, hoping they could help him.

Far from dismissing the Iceman, as Sanderson claimed they would, scientists rallied their forces to investigate this new discovery. After receiving his copy of Sanderson's memo, John Napier acquired a copy of Bernard Heuvelmans's article on the creature, just published (in French) in a Belgian journal.[60] Impressed by Heuvelmans's description and the fact it had been published in a scholarly periodical, Napier leaped into action. He went right to Dillon Ripley, by then secretary of the Smithsonian Institution, to see what could be done in order to

have the Smithsonian get the body. Ripley was well situated to deal
with the Iceman case. Years before, after mountaineer Eric Shipton
found the famous footprints in Nepal, he had contacted Ripley for
advice on how to search for the Yeti. Like many of the scientists who
had investigated the Yeti in the 1950s and been disappointed with
the evidence, Ripley still thought such creatures existed. He also now
had the weight of the United States federal government at his service.
Ripley, too, received a copy of Sanderson's memo and followed up on
it. They all then read Bernard Heuvelmans's article.

In the first few months of 1969 the halls of the Smithsonian
Institution fairly buzzed with excitement over Hansen's frozen body.
An informal Iceman Committee even came together.[61] Excited about
the possibilities of the find, William Charles Osman-Hill wrote Ripley
about doing something. He asked the Smithsonian Institution's secretary, "Can you bring any influence to bear on the problem, either
directly or through legal channels?"[62] Carleton Coon also contacted
Ripley to put his two cents in. He hoped the Smithsonian would
be able to get the body for proper study because "if it turns out
to be what it seems to be, this could be a major breakthrough."[63]
Professor Randy Hicks of Georgetown University wrote immediately
upon seeing the newspaper account. He told Napier, "I hope that the
Smithsonian will continue its efforts to secure or at least examine this
specimen."[64] On March 13, letters flew between various Smithsonian
departments to mobilize resources. Dillon Ripley wanted humani-
ties people as well as scientists to be part of the investigation, and he
contacted the legal department. General Council member Tom Jorley
contacted the FBI, just in case.[65] Ripley also went right to the source.
He sent a letter to Frank Hansen inquiring about access to the frozen
carcass by the Smithsonian. Hansen should allow this, Ripley said,
because the creature "may prove to be an outstanding contribution
to human knowledge."[66] To Ripley's surprise Hansen replied that he
no longer had the body. The "owner," he said, had taken it back, and
Hansen was not sure it would ever be available again. There would be,
however, a copy or "illusion" of the body put on display in the next
touring season.[67]

Napier contacted assistant secretary of the Smithsonian, Sidney
Galler, about acquiring the body. Sanderson suggested to Hansen
early on that the police might want to investigate. Napier and Galler
agreed such a threat might make Hansen a bit more pliable as far
as putting the thing into Smithsonian hands.[68] Hansen, growing
increasingly worried, told Ripley the creature was gone. Napier now
grew concerned about possible bad publicity for himself and the

Smithsonian if the thing disappeared. He also did not want to lose such a valuable prize if genuine. He put out a press release that, while stopping short of confirming the existence of a frozen ape-man, made it clear that the Smithsonian intended to investigate. Worriedly, he now told Ripley, "We have stuck our neck out by admitting to the press that we are interested [so that] I do not think that at this stage we can simply throw up our hands."[69]

Unknown at the time, Sanderson had been the first to contact the FBI shortly after he and Heuvelmans returned from Minnesota. Dillon Ripley now wrote directly to the director of the bureau, J. Edgar Hoover. He told Hoover that a scientific journal said the body was that of a human, and that it had been shot. After briefly looking into the matter, Hoover declined to investigate.[70] He told a surprised Ripley that the head of the Newark, New Jersey, field office had already informed Sanderson that unless some crime had been committed, the FBI could do little. Hoover finished by saying that if Ripley could find evidence a crime had been committed, the bureau would get involved.[71]

Drawing on the Hansen Case Memo, Sanderson wrote a popular article for dramatic release in *Argosy* and then a more scholarly one; meanwhile Heuvelmans, after quickly publishing his scholarly article, set to work on a popular book length treatment of the Iceman with the Russian author and Neanderthal relic theory supporter, Boris Porshnev. In *Argosy*, Sanderson asked with great flourish whether the creature "bridges the gap between man and ape." Calling the thing "Bozo," in the style of men's magazines, Sanderson relayed the great discovery as a breathless personal narrative. He began the adventure dramatically, saying, "I MUST [his capitals] admit that even I, who have spent most of my life in this search, am filled with wonder as I report the following!"[72] He then went on to recount his meeting with the creature in the barn. To support his contention that the Iceman was real he included quotes from the eggheads. Working from the photos and drawings Sanderson sent them, Carleton Coon, John Napier, William Charles Osman-Hill, and George Agogino gave the type of guarded reactions one would expect, neither condemning nor really endorsing the thing. Only Heuvelmans, who had seen it first hand, gave a rousing statement. "For the first time in history," he said, "the fresh corpse of a Neanderthal-like man has been found."[73]

To give readers a better look at the creature, Sanderson had the photos he had taken and drawings he had done converted into appropriate color magazine illustrations by science fiction and wildlife artist John Schoenherr. These "artist's interpretations" showed details

Sanderson believed the ice covering obscured from view on the magazine page. An experienced wildlife artist, Schoenherr made his first renderings chimplike. Sanderson rejected them because he wanted the creature to look more human. Schoenherr told Sanderson and the *Argosy* art director that the photos did not say to him human, but chimp. Determined to have a humanlike rendering, Sanderson insisted Schoenherr make the revisions. Needing the money more than he needed a fight with a past, and hopefully future, employer the artist relented and dutifully reworked the illustrations. The Iceman as the public would see him now looked more like a man than a chimp.[74]

Hansen's story of having a substitute dummy constructed undermined everything and put most of the scientific world off the story. Ever the optimist, Ivan Sanderson thought differently about the supposed substitution. The upstate New York newspaper, *Rochester Post-Bulletin*, ran a cover story in its April 21 issue with a picture of the "replacement" body. After careful examination of the photo, Sanderson sent Napier a typically long-winded letter describing why he thought the body was the same as the one he saw with Heuvelmans and not a substitute. He compared details of the refrigerator box and ice patterns on the glass. He felt Hansen had defrosted the original, repositioned it slightly, and then refroze it. "For these reasons," he told Napier, "I feel that Hansen's story [of creating a copy] is untrue."[75] Napier, however, had had enough. A few days after Sanderson's letter, Napier received a letter from a Missouri schoolteacher whose students had become interested in the Iceman because of Sanderson's *Argosy* article. The students enthusiastically discussed evolution in class and the Iceman's role and hoped the Smithsonian could clear a few things up.[76] Napier told the teacher that while the Smithsonian kept an "open mind" on the subject "the creature may well be a fake, indeed on the basis of probabilities this is the most likely explanation."[77] The next day Napier dispatched an official Smithsonian press release stating that the "Smithsonian Institution has withdrawn its interest" in the Iceman, and that it was likely a "carnival exhibit" not a living creature.[78] Napier was not the only one in Washington who had grown tired of the charade. Around this time the Smithsonian curator of anthropology, Lawrence Angel, made it clear the Smithsonian had had enough monster nonsense and would no longer be researching or collecting such materials.[79] Angel had studied at Harvard under Earnest Hooton and Carleton Coon. A specialist in human evolution and diversity, he had the proper pedigree to be an ABSM believer, but the claims and counterclaims by the crackpots and the lack of credible evidence had put him off as well.[80]

While he still thought the Iceman genuine, Sanderson also appreciated the baggage the creature carried that made the carcass problematic. He told his old friend Ralph Izzard that "the whole thing has got so mixed up with carnivals, plastic copies made in California, smuggling and Christ knows what else that we cannot be definite about anything." He added cryptically that "there's a heck of a lot more to this story but I am not yet at liberty to publish on it."[81] Hansen, who by now had changed his story so many times even he could not keep track of it, had one more thing to add. At Sanderson's prodding, Hansen wrote an article (likely with Sanderson's help) explaining his side of the story. The dramatically titled "I Killed the Ape-Man Creature of Whiteface" only seemed to muddy the waters further.[82] Hansen claimed he had not faked anything and feared for his freedom now that the Customs Department and Department of Health, Education, and Welfare were supposedly nosing around (thanks to Sanderson having alerted them to the creature's existence). Despite the FBI's documented lack of interest, Hansen claimed he feared the Bureau, too. Finishing his version of the tale, Hansen said skeptics would claim the whole thing a hoax anyway. He then added with the voice of a barker, a classic carnival flourish: "Possibly it is, I am not under oath." Reading the last paragraph of Hansen's article, one can almost hear the sounds of the midway in the background. John Napier, who by now thought the Iceman a fake, said of Hansen's article, "I don't believe a word of it."[83]

The relic proponents lost no time getting their views out in print, with the Iceman as the central artifact. With his popular article getting attention on the newsstands, Sanderson now turned to the scholarly community. Sanderson wrote once again for the Italian journal *Genus*.[84] This article ran somewhat longer, but toned down the drama of the *Argosy* piece and the Hansen Case Memo. For a "scholarly" article, however, it contains no footnotes or citations. In collaboration with the Russian mystery-ape investigator Boris Porshnev, Bernard Heuvelmans released a popular book length work. *L'Homme de Néanderthal est Toujours Vivant* (1974) summarized all Heuvelmans's work on the Iceman, particularly his speculations on it being a Neanderthal relic (Sanderson thought it more likely the Iceman was a *Homo erectus* relic), the *"des hominoides reliques"* concept. It opened with a survey of the various legends, stories, and evidence for archaic hominid survivals from around the world. He then presented his work on the Minnesota Iceman, including a group of photos he had taken in the cold, dark trailer with Ivan Sanderson. While Heuvelmans thought the Yeti possibly a *Gigantopithecus*, he

considered the Iceman a Neanderthal. Heuvelmans and Porshnev
made a passing reference to a much earlier study on relic Neanderthals
by Polish anthropologist Kazimierz Stolyhwo (1894–1976), who
claimed in 1908 to have unearthed a Neanderthal burial that contained iron body armor.[85] Stolyhwo did careful work comparing the
skull cap he found to known Neanderthals to support his contention
that this particular Neanderthal at least survived into the modern
era. Most anthropologists then and now felt he misidentified a robust
modern human of the tenth century for a caveman. Heuvelmans and
Porchnev may have had their doubts as well, as they did not make
much of Stolywho's claims. In 1699, beating them all to the punch,
British anatomist Edward Tyson (1650–1708) published his groundbreaking treatise on primate anatomy. In the introduction to his
*Anatomy of a Pygmie*, Tyson explained that the motivation for writing his book was to prove that legends of satyrs and pygmies from
such authors as Pliny, Aristotle, and Strabo described apes rather than
some unusual form of human or monster.[86] Heuvelmans, Porshnev,
and especially Sanderson, discounted Tyson's work and insisted classical references to such manlike monsters confirmed there were surviving Neanderthals. American Bigfoot enthusiasts knew of *L'Homme
de Néanderthal*, but few of them read French, so Sanderson's articles
had more direct impact on their research.[87]

In June of 1969 John Napier wrote his boss Dillon Ripley a wrap
up of the Iceman incident. He told the secretary of the Smithsonian
Institution that only one Iceman body had ever existed. Unlike
Sanderson, who also believed there had only ever been one body—a
real one—Napier thought the body faked. "I am now satisfied," he
said, "that the Iceman is no more real than any facsimile in any wax
museum."[88] Despite his initial enthusiasm Napier admitted they had
been led around by heightened expectations on their part. Hansen
never intended to fool experts, he thought, but simply to promote his
sideshow, but things got out of hand. Napier and the others had also
been brought around by the glowing reports of the thing, reports
produced by Sanderson and Heuvelmans. Napier questioned "how
two experienced and capable observers could have been taken in?"
He was less willing to acknowledge how easily he and other academics had been swayed by Sanderson and Heuvelmans. Willing to give
the Scottish naturalist and the French zoologist the benefit of the
doubt, Napier did not chastise them too much. He understood that
for years the two men had been looking for just such a thing. Shortly
after, when Napier wrote his version of the Iceman story for his book
on Bigfoot, the Iceman chapter was originally meant to be titled

"The Faultless Monster," meaning it could satisfy many an observer's view of what such a creature should look like.[89] When Sanderson and Heuvelmans unexpectedly found themselves confronted by a well made and presented fake, it was easy to understand how their excitement and desires could get the better of them. Training could sometimes be blinded by desire. Both Heuvelmans and Sanderson loved to argue how scientists refused to accept their findings. They refused to acknowledge how easily—in the Minnesota Iceman case—quickly, and eagerly those same scientists had followed them down the rabbit hole.

## Conclusion

The Minnesota Iceman case came close to ending scientific interest in anomalous primates forever. After such a public debacle, few scientists, even if they thought the creatures real, would be likely to go public now. Then, a young man from Utah saved these storied creatures from complete marginalization. Having narrowly missed his own personal oblivion, Grover Krantz stepped into the position as one of the few scientists to openly investigate anomalous primates by trying to work out how Sasquatch fit into an evolutionary context. John Bindernagel, Igor Bourtsev, and a few others also followed the trail but did not yet have a public face. Because of the particular circumstances of his life, Krantz (who had just missed seeing the original Iceman at a carnival) managed to devote a good part of his career to studying the phenomena. As George Agogino pointed out in the 1950s, however, an academic took a great risk with their career by openly pursuing these creatures. Grover Krantz took that chance. He knew his career could be ruined and his advancement cut off, leaving him to live in a broken-down trailer for the rest of his life, like his nemesis René Dahinden. Despite resistance and hostility from within the ranks of the mainstream, and threats and ridicule from the fringe, he kept on. Krantz's career is a case study in the murky borders between science and pseudoscience, and is a microcosm of the search for monsters.

# Chapter 4

# The Life of Grover Krantz

*And what is not regarded as wondrous when it first gains public attention? How many things are judged impossible before they actually happen?*

Pliny the Elder, 77 AD

Grover Krantz occupied a unique and quirky place between the crackpots and the eggheads. A trained paleoanthropologist, by the middle of the 1970s he had come to the realization that for him straightforward evolution studies held little opportunity for notoriety. He therefore determined to win renown through his work on manlike monsters. He knew this would be a risky course to pursue. Fulfilling his expectations, Krantz's career suffered, but the role of maverick appealed to him. Such a position allowed him simultaneously to support and criticize both sides of the issue and situate himself as the academic authority and leader in the field. He enjoyed taking the road less traveled, which he did from conviction, but also from the simple joy of contrariness. He would have opted for such a trajectory whatever area he devoted himself to, whether the lives of Neanderthals, the evolutionary patterns of *Homo erectus*, the migration histories of Indo-European peoples, or manlike monsters. He entered the latter field just as the Yeti/Asia phase ended and the Sasquatch/North America phase took off, finding himself with a tortuous route to negotiate between competing camps in an attempt to reach his goals. If any academic had the motivation to step into the public eye and make a career of supporting the existence of anomalous primates, it would have to be someone like Grover Krantz.

## The Poor Scholar

Born in Salt Lake City, Utah, in 1931 Gordon "Grover" Sanders Krantz came from hardy Mormon stock. His parents, Victor Emanuel and Ester Marie Sanders Krantz, were born in Sweden and immigrated to Chicago, finally settling in Utah. They came from a line of pious members of the Church of Jesus Christ of Latter-day Saints (LDS), better known as Mormons, but Grover showed little interest in the religion. Seeing her son drifting further and further from the fold of the faithful, Krantz's mother worriedly questioned him on the subject. She saw his growing interest in geography and science, and loss of interest in religion. Attempting to explain his theological beliefs to her in 1952, he said he rejected the notion that people should humble themselves before God or that God should be worshipped as the creator of all things. He told her that while he tried to follow a basic Christian philosophy of behavior and morality, he favored logic and reason over superstition and dogma.[1] As a teen he already exhibited the tendency to go against the mainstream, which would be the hallmark of his professional career.

Strangely, Bigfoot already stood in proximity to Krantz's life in the form of a Mormon connection. Within LDS lore early church Apostle David W. Patten claimed that in 1835 he had an encounter with a strange being in Tennessee. Patten told the story of a tall, hairy "man" who accosted him along a lonely back road and engaged him in a discussion about the state of the soul. Patten concluded the hairy creature must have been the Biblical Cain. The story became popular within LDS circles and survives to this day. Some later Mormons and others suggested that Patten actually encountered a Sasquatch.[2] It is unclear whether Krantz knew the story. He certainly never mentioned it in any of his writings or correspondence. If he did know of it as a young man, he likely dismissed it as superstition. His youthful worldview yearned for the rational and the concrete. He preferred mammal skeletons and their function to bogeymen meant to scare people into right behavior. In his teens he eagerly collected any cat or mouse carcass he came across and recovered their skeletons. As a college freshman he wrote in his diary with enthusiasm, "Beheaded a cat, then ate BIG [his caps] turkey dinner."[3] For years he carried around with him a tiny black and white photo of a mouse's entrails.[4]

In 1949 he entered the University of Utah as an anthropology major, but the next year joined the Air National Guard as an enlisted man , serving with the 191st Fighter Bomber Squadron (his older brother, Victor, served as an officer in the Air Corps and a bombing instructor). Honorably discharged in 1953 as an airman second class, Grover Krantz left the land of the Mormons forever, transferred to the University of California, Berkeley, and resumed his formal education.[5]

The anthropology department at Berkeley had been established in 1901 with funds from Phoebe Hearst, mother of newspaper magnate William Randolph Hearst and a passionate and prodigious collector of Native American artifacts. The department's first professor was Alfred Kroeber (1876–1960), the influential mentor to the nascent anthropological community in America. Born in New Jersey, Kroeber found Native American culture fascinating. He argued that American science in general, and archaeology in particular, had to move beyond the amateur tradition of simply classifying artifacts and into placing them in a cultural context in order to understand them. Only the second American to be awarded a doctorate in anthropology, he studied at Columbia University, New York, under the tutelage of the legendary ethnologist Franz Boaz. Since its inception, the Berkeley anthropology department, as did the school's other biology-related departments, produced a virtual army of researchers and has had on its faculty scientists who have made many of the great discoveries and breakthroughs in the field. Krantz became part of a powerful American scientific tradition and imbibed ideas from many quarters.

The year he left the Air Force, Krantz, now 21 years old, married 17-year-old Patricia Howland in a ceremony in Wyoming. They headed to California, but married college student life and a constant lack of money put a strain on their relationship, and they divorced in 1958. Krantz seemed to enjoy married life, or at least felt a need to have a companion, because he soon married again. He had four wives over the course of his life.[6]

Krantz's university student years were dotted with signs of great professional promise and with as many signs of personal disaster. After transferring to Berkeley, to make ends meet he worked for a year as a curator at the San Mateo Junior Museum and then as a social worker for Almeida County. Neither job paid particularly well. He did not do well academically his first few semesters, either. While his anthropology grades fell in the A and B range he did poorly in languages (getting a D in both German and Slavic, and a C in French). He persisted despite the personal hardships, raised his grade point average, and graduated in 1955 with a degree in anthropology.[7] He then entered the anthropology graduate program, where his grades improved markedly. As a master's degree student under Theodore D.

McCown (with whom he studied as an undergraduate) and others, he worked as a teaching assistant in the anthropology department. McCown (1908–69), a scientist of international reputation, opened up the important Neanderthal cave sites in Israel at Mount Carmel with British anthropologist Arthur Keith in the 1930s. Though prestigious, the teaching assistant position with McCown did not pay enough to live on, so from 1960 to 1966 Krantz had no choice but to stretch out his graduate studies by not immediately entering the doctoral program after finishing the master's degree. He augmented his meager wages back in museum work as an exhibits preparation technician at Berkeley's R. H. Lowie Museum of Anthropology.[8] All this time he saved his pennies and dreamed of reentering academic science. During this period away from formal studies he still managed to publish a number of short scholarly articles, including "Sphenoidal Angle and Brain Size," in the journal *American Anthropologist*.[9] His earnings being what they were, he slipped into genteel poverty. Broke, divorced again, living in a tiny, rundown apartment, he borrowed money from relatives just to survive. He modified his clunky, but relatively reliable car by tearing out the back seat and installing a platform for a bed that extended into the trunk. Now 32 years old, in spite of some minor successes, his career had stalled, and he began drinking heavily.[10]

## Dog Saves Man

Krantz decided in the midst of all this turmoil that he needed a dog. His depression caused him to drink and lose the few friends he had. "My life at that time," he later wrote, "consisted of a part-time job and nearly full-time drinking."[11] He had always loved dogs and found them better companions than most people. He especially liked the big, lanky varieties and dreamed of someday getting an Irish Wolfhound. He once tried to buy one through the mail as an undergraduate, but when the crate arrived with a puppy in it, the creature turned out to be a Greyhound. Nebby (short for Nebuchadnezzar) turned out to be a good friend anyway, though he had a penchant for getting roughed up by a group of Great Danes that apparently roamed the Berkeley campus at will, accosting passersby, both bipedal and quadruped. In a way telling of his personality, Krantz vowed to get an Irish Wolfhound "that would clean up those Danes."[12] Then in 1963 he saw an ad and felt the time had come to get the real thing. When he arrived at the kennel his shabby appearance worried the kennel owners, making them reluctant to sell him a puppy for fear he could not afford to care for it. After some haggling, he convinced them and left with an Irish wolfhound he named Clyde. This animal saved Krantz's life. His spirits picked up, his drinking lessened, and he and Clyde became a common sight on the Berkeley campus. The dog did not tax him the way humans did. Clyde always came when called, never questioned him, never made him feel inadequate, never let him down. Krantz came to rely on Clyde to screen potential human relationships (the dog took a particular liking to a young woman named Evelyn, who became the third Mrs. Krantz). He also entered Clyde in a series of dog shows, where the animal exhibited potential star quality by winning several blue ribbons.[13]

As Clyde won awards, his owner decided to take a trip out to the woods. Grover Krantz's interest in manlike monsters had begun when he had read reports of the Yeti of Nepal in the 1950s. He had come across these reports as an undergraduate as he pondered the idea that Neanderthals had crossed the Bering land bridge into North America. He later told Bigfoot hunter Roger Patterson that "in 1953 I first heard reports about a giant manlike creature in Western North America, and I have been gradually accumulating information on the subject ever since."[14] That same year he executed a series of detailed anatomical drawings of various hominid fossil skulls as part of his class work, which included one of *Gigantopithecus*, the fossil that would dominate his theoretical work.[15] Bigfoot stories from California intrigued him, as did the literature of the emerging field of cryptozoology. First he read Ivan Sanderson's *Abominable Snowmen* and then Bernard Heuvelmans's *On the Track of Unknown Animals*. After digesting these works Krantz initially thought the Yeti might be a Neanderthal or *Australopithecus* survival, as did Heuvelmans and Sanderson, but then began to consider the possibility of it being a living relative of the fossil primate *Gigantopithecus*, familiar to him through his course studies of the work of Ralph von Koenigswald and Franz Weidenreich.[16]

In 1964 Krantz went to Bluff Creek, where reputed Bigfoot tracks had been famously discovered in 1958. When the original discovery was made, his professor, Theodore McCown, had publicly commented on the finds that had caused such a commotion. Showing some familiarity with the topic, McCown considered such tracks as an "old story" and that if the creature turned up it would solve the mystery that had perplexed the Pacific Northwest for years.[17] Though Krantz came back disappointed to have found nothing at Bluff Creek, the idea of a big hairy apelike creature running around the Americas stuck with him.

In the mid-1960s Krantz began to feel better personally. Clyde and Eve buoyed his depression; he got control of his drinking, finally entered the Berkeley doctoral program—working with Sherwood Washburn—and began to pick up steam. This put him in the middle of one of the legendary conflicts of postwar American anthropology and the question of race. The experience left a mark on his research and had major implications for the course of his career. Known to his friends as "Sherry," Sherwood Washburn (1911–2000) pioneered primatology and anthropology studies at Berkeley. His work on primates in the 1940s and 1950s, done in the wilds of Kenya rather than on creatures in captivity, blazed new trails. He taught anthropology at Berkeley from 1958 until 1978 alongside Theodore McCown, who had reached the end of his long career as Washburn hit his stride. Washburn studied under Ernest Hooton and Carleton Coon at Harvard, but largely rejected their views on race and society. Instead, Washburn worked to bring evolutionary thinking into anthropology and coined the term "The New Physical Anthropology." Washburn focused on teaching his students, including Krantz, the important interrelated nature of evolution, skeletal anatomy and musculature, and human and primate social behavior as well as human origins and ethnic studies. His classes often concluded with standing ovations from students. He argued that tool use, the division of labor between male and female, and especially hunting techniques, influenced human evolution.[18] Krantz's published work would reflect a good bit of Washburn's teachings. Their personal relationship would be a rocky one, however.

In 1965 Krantz received an offer to write a short article on anthropology for *Scientific American*. Thrilled by the opportunity to gain such notice, he went to Washburn for advice. Rather than welcoming Krantz's opportunity, Washburn told him that no graduate student had the skills to write for such a prestigious publication. Washburn then used his contacts to scuttle the offer from the magazine. The row that ensued led Washburn to also suggest to Berkeley officials that Krantz should not be allowed to finish his doctoral program. Just then a temporary position opened at the University of Minnesota to fill anthropologist Rupert Murrill's sabbatical absence. Krantz desperately applied, was accepted, and happily left the titan of anthropology behind before he did more damage to his career.[19]

The sabbatical fill-in position at the University of Minnesota worked out better than Krantz could ever have expected. He generated such high regard among faculty and students that he was asked to stay on a second year despite the return of Professor Murrill.

The department chair, E. Adamson Hoebel (1906–93), troubled by Washburn's behavior, recommended Krantz finish his doctoral studies in Minnesota rather than return to Berkeley. Hoebel worked on the legal practices of Native Americans. He was heavily influenced by Karl Llewellyn and the sociological legal realism school of thought at Columbia. With an atmosphere so welcoming and supportive, Krantz agreed to stay. He quickly took, and passed, his qualifying examinations and started writing his dissertation on human evolution.[20]

In October 1967, as Krantz grew comfortable as a professor, Roger Patterson and Robert Gimlin wended their way slowly on horseback through the woods of California and had their date with destiny. Though there had been a few newspaper stories about their capture of Bigfoot on film, the cover story on the incident, published in *Argosy* magazine's February 1968 issue and written by Ivan Sanderson, grabbed Krantz's attention.[21] *Argosy* included several stills from the film on the cover and in the article. Bigfoot now had a face—albeit a dark and blurry one—to go with the legend. In Minnesota someone heard that visiting professor Krantz had gone looking for the Snowman in California, so the *Minneapolis Star* asked him for a comment on the *Argosy* pictures. Krantz now made his first public statement on anomalous primates. Unimpressed by what he saw in Sanderson's article, he said the pictures "looked to me like someone wearing a gorilla suit."[22]

Part way through Krantz's second year at Minnesota Hoebel received word from Washington State University anthropology department chair, Robert Littlewood, about suggesting someone for a full-time position that had just become available there. Hoebel immediately put Krantz forward, saying Krantz "is too good to be lost to the academic profession."[23] Krantz immediately wrote to Littlewood, applying for the position. Littlewood wrote across the top of Krantz's letter "Interesting-broad interests."[24] Another Minnesota faculty member told Littlewood that Krantz's work "can only rebound to the credit of our department."[25] A few months later, Krantz had the job. Well along to finishing his doctoral dissertation in September of 1968, Krantz stepped confidently into his new position as assistant professor of anthropology and settled in for a 30-year career at WSU. He set to work completing his graduate thesis and, in 1971, the University of Minnesota awarded his doctorate in anthropology with a concentration in human evolution.[26] His life and career had taken a decided turn for the better and besides Irish wolfhounds Krantz's interests turned more and more to another hairy animal.

## The Bossburg Incident

With a general enthusiasm for the topic, Krantz wavered as to whether manlike monsters actually existed. His position began to change after Roger Patterson shot his jumpy film. At first skeptical of the stills in *Argosy*, after he saw the actual Patterson film in 1969, the realism of the creature's locomotion impressed him. He revised his initial assessment and began to consider that the creature might be real. He also began to delve more deeply into the field, moving from simple curiosity to serious research.

A few months after Krantz saw the Patterson film, he read Canadian Sasquatch enthusiast and journalist John Green's *On the Track of Sasquatch* (1968), with its compendium of eyewitness and historical accounts of the creature. Green said, "The origin of such a creature is no problem."[27] In 1946, while in university, Green took an anthropology course where he learned of *Gigantopithecus*.[28] It seemed an obvious step for Green, once he began to encounter Sasquatch stories, to bring the Asian primate into the discussion despite having missed both Carleton Coon's and Bernard Heuvelmans's references to it. Green argued that evidence for such a creature living in Asia made it probable that one might be in North America. He said the description of *Gigantopithecus* was "a pretty good thumbnail description of what people have been seeing all along."[29] Green said the creature must have made it into the Americas along the Bering land bridge. He also made reference to Theodore McCown's remarks about the Bluff Creek tracks. While impressed by Green's book, Krantz needed something more concrete.

During the Thanksgiving holiday of 1969 the event that changed Krantz's career occurred. Krantz saw Sasquatch tracks in the field for the first time at Bossburg, Washington.[30] The Bossburg incident, maybe more than any other in the pantheon of Sasquatch events, came as a watershed for anomalous primate research. It marked the point at which an American academic scientist began to take the Sasquatch phenomenon seriously and pursue it actively and publicly over an extended period (George Agogino and Carleton Coon made little of their work public, John Bindernagel was not yet well known and John Napier did not make it the focus of his career). At first blush the Bossburg case seemed to everyone involved to be the one to crack the Sasquatch question wide open. It began with an energy that had all involved especially attuned to the possibilities of their work finally being vindicated.

The general area around Bossburg had been searched the previous spring. By fall the hunting season opened, so searchers naturally felt reluctant to go into the area with armed men about. Norm Davis, operator of the local KVCL radio, claimed a Vancouver journalist contacted his wife Carol on November 20th and told her Ivan Marx had sighted tracks in the Bossburg/Colville area. Butcher Joe Rhodes of Colville suggested to Marx that he should keep his eyes open for tracks when he went out hunting because of a woman's report of the creature the previous spring. (Rhodes supplied meat to Marx's cougar farm). Marx went out and indeed found tracks near the municipal dump. Marx agreed to take the Davises and Joe Rhodes to the dump to see the tracks for themselves. The group made casts and took photographs. A pair of dumbfounded state wildlife agents, Wayne Wendt and Ron Trim, arrived on the scene and said they had never seen such tracks before. John Green then called René Dahinden. Green's previous commitments prevented him from coming right away. Dahinden pondered the account for a few days then proceeded to Colville.[31]

Some months after the hubbub at Bossburg died down, Marx came up with film footage of the creature. He staked out various spots in the area and, after some patient waiting and woodland tracking, he filmed it. The footage showed a dark, hairy, though not overly large or bulky creature limping and favoring an injured arm as it scampered about a meadow. Peter Byrne had by now arrived on the scene and made a deal with Marx to put a copy of the film into a safe-deposit box and Marx on paid retainer. Byrne then did some investigating. With the help of local children he located the spot in the film and found, by comparing known objects seen in the film, that the creature was only about six feet tall, not eight or nine as Marx claimed. Byrne then discovered blank film in the deposit box. Also, the original film itself, when people saw it, was obviously—even laughably—a man in an ape suit. Marx's claims fell apart as it became clear his film was a fake. In his assessment of the Bossburg case, anthropologist David Daegling, quipped that "a confidence trick is still a trick, even if it is hatched at the town dump."[32] After the smoke cleared, some of the crackpots—though not John Green—walked away from Bossburg feeling it another disappointment. The strange tracks, while long since melted, simply refused to go away.

## The Cripplefoot

The Bossburg case proved to be a major turning point in the life of Grover Krantz. The typical trajectory of anomalous primate discoveries has the crackpots becoming excited about a discovery, feeling it to be genuine proof of their assertions. The eggheads come along and, after the occasional mild interest, usually dismiss the whole thing as a hoax or misidentification and walk away. With Bossburg, the opposite occurred. Here the eggheads became excited while many crackpots eventually dismissed it. Grover Krantz arrived on the scene late, but his attention immediately turned to the abnormal right foot prints.

As Bossburg was relatively close to the WSU campus at Pullman, Washington, and the school term had ended for the holidays Krantz headed for the Canadian border. He arrived in mid-December, when most of the tracks had already been destroyed by nature, by plaster casting, or by the army of gawkers who had stomped about. Once there, Krantz met John Green who had only just arrived himself. Krantz admired Green's book, intrigued by his use of *Gigantopithecus*. Green liked the idea of finally finding an academic sympathetic to his ideas. Luckily, someone had protected one of the better tracks by covering it with a cardboard box and newspapers.[33] The anthropologist and the journalist went to see it. Finding the spot and carefully removing the protective covering, the anthropologist leaned down to examine the first live Sasquatch track he had ever seen. As Krantz squatted there in the snow with his wool cap and rubber boots, staring into a foot-shaped destiny, Green took out his camera and snapped a picture, recording the moment for posterity. Krantz may have not realized it at that moment, but he had just turned permanently from skeptic to believer. In that moment an obsessive career he had never dreamed of began in earnest.

Krantz made a cast of the print, but later acquired an even better set of the cripple tracks, made from some of the first and best tracks found. These peculiar prints stamped themselves upon his mind and would not go away. The summer following the Bossburg incident, Ivan Marx found several alleged Sasquatch handprints left in mud. Marx, whose star had yet to fall, made plaster copies of the prints and loaned them to Krantz for study. Drawing upon his experience in museum work, Krantz first made latex copies of the plaster casts to work from. He now had artifacts he could study in conventional anthropological terms. The handprints (both left hands), had unusual details that caught Krantz's attention. They made him think he might be able to produce a scholarly article on them. His examination of the casts and the details he found suggested the prints came from a large primate.[34] The sheer size of the prints impressed him, but one little thing was crucial. He noticed that the thenar eminence, the pad of muscle just below the thumb, was greatly reduced. The thenar pad pulls the thumb inward and gives humans an opposable thumb. A tiny thenar eminence meant no opposable thumb, which meant no humanlike grasping hand, and reduced the likelihood of hoaxing. He wrote up his findings and submitted them for publication. Krantz's enthusiasm was tempered by the fact that none of the ten scientific journals he submitted his paper to accepted it for publication. He found it difficult to believe this topic should not be discussed in a scholarly forum. After all, he made no wild claims or outlandish statements, but examined a natural phenomenon in sober scientific terms. Krantz's Sasquatch career may have ended right there, if not for John Green.[35]

### Publishing Bigfoot

Excited by the activities of 1969, including the Bossburg incident, John Green published his second book, *Year of the Sasquatch* (1970). Similar to his first, but focused on one year only, Green recapped all the sightings of the creature as well as related news from around the world. Long since a veteran of the cool response of the scientific community, Green saw a bit of hope in the goings on in Soviet Russia. Boris Porshnev, the Russian historian and psychoanalyst, had established himself as the premier anomalous primate investigator of the Soviet bloc, where such creatures, called Almasti, had a long history. He encountered the same sort of resistance there that North American researchers had. However, as Green enthusiastically pointed out, the *Journal of Soviet Ethnography* had published one of Porshnev's Almasti-related articles.[36] The journal explained that while the editors did not agree with Porshnev's findings, they felt obliged to run the article so that more scholars could see this work and comment on it. "One could wish," Green said, "that there were scientific publications in North America with editors who took a similar attitude."[37]

To his surprise, Green's modest call received an answer. It came from anthropologist Roderick Sprague of the University of Idaho, who edited *Northwest Anthropological Research Notes* (NARN), a scholarly journal specializing in the anthropology of the Pacific Northwest. Sprague was familiar with the many reports of manlike creatures as well as with Native American folklore of hairy monsters. Like his Russian counterparts, he was unimpressed with the evidence for anomalous primates, but Green's books and call for scholarly publication did. As a result Sprague included an editorial in the fall 1970 issue of NARN, calling on academics to submit appropriate papers on Sasquatch. Sprague justified his actions by saying anthropologists "have an obligation to study the sociological and anthropological implications of the belief systems which contain or encourage the continuation of such beliefs" as the existence of anomalous primates.[38] One of the first to submit a paper was Grover Krantz, whose flagging spirits had been raised by Sprague's editorial. Sprague and his editorial board liked Krantz's paper on Sasquatch hands and ran it the following fall. This would be the first in a series of articles on Sasquatch written by Krantz that would be printed in NARN. At the same time, however, NARN would be the only science publication—except for the journal of the International Society of Cryptozoology—to publish any of Krantz's scholarly Bigfoot writings.[39]

Krantz's article on Sasquatch handprints was not the first such attempt at anatomical reconstruction of an anomalous primate artifact. In 1960 Wladimir Tschernezky, an anthropologist working in London, published a short article on Yeti tracks in the prestigious journal *Nature*. Anticipating some of Krantz's work, Tschernezsky looked at Eric Shipton's famous Mt. Everest Yeti photograph and rebuilt the foot that made it. He based his reconstruction on the photo, then made his own footprints in sand and compared them to the footprints of other animals. He also considered whether a deformed human foot had made the supposed Snowman print. His analysis led him to believe that Snowman walked in the same manner as a human. In a paper Krantz must have missed (he makes no reference to it in his own work), Tschernezsky concluded that "all the evidence therefore suggests that the so-called 'Snowman' is a very huge, heavily built bipedal primate." Then he suggested it was "most probably of a similar type to the fossil *Gigantopithecus.*"[40]

In early 1970, prior to the publication of the Sasquatch handprint article, Krantz received a good plaster-cast set of the Bossburg Cripplefoot tracks. He pressed these tracks into "a prepared bed of fine dirt" so that further good quality copies could be made. Krantz made latex copies from this second generation cast.[41] Using his training as an anatomist he reconstructed the structure he saw in the track and generated a drawing of the creature's foot bones for both the normal left and crippled right foot. He compared this drawing to a drawing of a human foot skeleton blown up to the same size as the cripple. Then he did a series of calculations working out the biomechanics of a foot that would be needed to support an eight-foot-tall biped a third larger than the average human. He worked out how big the heel should be, where the ankle should attach, and so on. He then compared the calculations to his drawing. He anticipated the theoretical structure would not match the skeletal drawing. This would show up the Cripplefoot cast as a hoax. To his surprise the theoretical calculations and the drawing matched. He did not have to finesse the data, they matched immediately. The Cripplefoot was built just the way biomechanical theory said it should be.[42] The anthropological approach had begun to produce results.

## The Issue of Typology

Concerning his methodology, the most influential scientist in Krantz's life was not Theodore McCown or Sherwood Washburn, but anthropologist and secretive Yeti enthusiast Carleton Coon. While Krantz never studied with Coon, he read Coon's work closely and they eventually became friends. Coon took a typological approach to race and evolution. A common technique in the early stages of the history of physical anthropology, typology and somatotyping (body typing) compared humans by measuring skeletons, body parts, heads, and other details to see the similarities and differences. Such data could be perverted to support the idea that these differences indicated the relative inferiority or superiority of different groups. Nazi race theorists made extensive use of typology in the 1930s and 1940s.[43] Coon's colleague at Harvard, William Sheldon, used such techniques to study the physiques of college students to study racial difference. He coined the still-popular terms *endomorph*, *mesomorph*, and *ectomorph*, to categorize basic body shapes. His book on human body types based upon this research, *The Atlas of Men* (1954), became an important source for Krantz's study of Sasquatch body types. Sheldon pursued questions as to whether body type had any relation to temperament, intelligence, criminality, poverty, or other social behaviors or conditions. While Krantz rejected the racist aspects of typology in favor of the view that ethnicity appeared as the result of adaptation to local environmental conditions, he did employ typological techniques of data acquisition in order to build a biomechanical image of Bigfoot.[44]

By the mid-twentieth century Darwinian natural selection found itself being used in a tug of war between conservative political and segregationist forces and antiracists and progressives. Carleton Coon fully embraced typology as a way to determine the basis of racial and ethnic difference: unfortunately, so had the Nazis Coon vigorously and heroically fought against during the war. The strong reaction against the Nazi eugenics program had a profound effect upon anthropological theory and practice as progressive and liberal antiracists worked with great passion to scrub their profession of anything they felt harked back to the pseudoscience of the Third Reich, America's slavery past, or the growing violence of the anti–civil rights backlash. Coon said he wanted to answer questions of human variability, not determine superiority or inferiority, and typology represented his favorite approach. He spent his career traveling the globe—especially the Middle East—measuring local people's anatomy. This helps account for his interest in Yeti and Bigfoot. Manlike monsters represented one more exotic group to measure to explain human types. Unfortunately for him, American anthropology increasingly equated typology with pseudoscience. Typology had also fallen out of favor because it produced little of scientific use as further research showed physical stature had little to do with intelligence. It likely would have withered on the vine even without its dark past.[45]

Along with the growing acrimony over race, another problem dogging anthropology in the middle of the twentieth century centered on adapting evolutionary theory. This situation arose in part because even in the 1960s many anthropologists still thought in terms of individual organisms rather than populations the way evolutionary biologists did (monster hunters tended to do the same). Coon's attempt to combine evolution theory with anthropology mimicked a late form of an older idea termed *polygenesis*. This idea, popular in the nineteenth century, held that each human race had a separate biological, and even theological, origin. This made them separate species, and opened them up to being categorized as superior or inferior. This kind of thinking angered men like Sherwood Washburn, Theodosius Dobzhansky, and Ashley Montagu, who took a more progressive attitude to race.[46] While Coon's work stood separate from polygenesis, the climate was such that comparisons came easily.

In his controversial human evolution writings, Coon argued that the modern human races resulted from *Homo erectus* populations adapting to their environments separately due to different pressures. This resulted in each modified *erectus* group evolving into a separate subspecies of *Homo sapiens*. In other words separate *Homo erectus* populations around the world evolved independently into five different forms of *Homo sapiens* as represented by the five modern races.[47] This left open the door for racialists and segregationists to argue that African Americans and other people of color stood separate from, and biologically inferior to, Caucasians.

Coon's work would have been seen as provocative on its own, but when it became the scientific support for openly racist authors, a storm broke. Its epicenter was *Race and Reason* (1961) by former Delta Airlines executive, amateur anthropologist, and white supremacist

Carleton Putnam (1901–98). Coon's *Origins of the Races* figured prominently in Putnam's book as the scientific source for his social assertions about the necessity of school segregation and the low quality of Black intelligence.[48] Prior to the publication of this book, Coon had corresponded positively with Putnam, who was his cousin, discussing race issues and the future, all of which made Coon seem more complicit.[49] Coon insisted he was not liable for how others interpreted and used his work, nor was it his responsibility to publicly or privately oppose such use. The backlash against Putnam redounded fully and disastrously upon Coon. This public controversy resulted in Coon's career being trashed and his taking early retirement from Harvard. Paleoanthropologist turned historian of science, Pat Shipman, called Coon "a man betrayed by history."[50]

The anthropological community reacted strongly to what it perceived as an attempt to hijack science for nefarious ends. Grover Krantz's doctoral advisor, Sherwood Washburn, stepped forward to lead the antiracist charge. He felt, along with influential evolution theorist Theodosius Dobzhansky, that Coon was disingenuous about not having anything to say about how segregationists used his work. Like Coon, Washburn studied under Earnest Hooton at Harvard. He rejected Hooton's racialism and methods, including the kind of typology and somatotyping both Coon and Krantz embraced. Washburn argued "typology must be completely removed from our thinking."[51] He considered the old typological approach outdated, useless, and misleading. The proper avenue for studying human variation, Washburn argued, originated in genetics. He denounced Coon publicly at a meeting of the American Anthropological Association on November 16, 1962. Dobzhansky added that while Coon's work might be "attractive to racist pamphleteers," like Putnam, it was not to evolutionary biologists.[52] In *Mankind Evolving* (1962) Dobzhansky said that races did exist, as did racial difference, but they are so intermixed as to make the concept meaningless. He argued humans interbred so much that pure races did not exist, if they ever did, and that the close proximity of cultures causes races to converge more than biology might have them diverge. Dobzhansky participated, along with Ashley Montagu, in the production of the United Nations Educational, Scientific and Cultural Organization (UNESCO) statement on race published in 1950. Produced as reaction to World War II, the UNESCO statement denounced race theory. Leaning toward the progressive end of the political spectrum, Krantz attempted to distance himself from racists by stating: "a race is a portion of the species that is distinguished by inherited biological traits and has (or

had) some geographical concentration." This definition, he continued, "rules out any social concept of race based on nationality, religion or language."[53]

Krantz had studied with Washburn during the strife ridden
years of the late 1960s and watched as the colossi of anthropology
maneuvered and jousted on the field around him. Krantz's sympathies, however, fell on the conservative Coon rather than the liberal
Washburn. Krantz and Coon found in each other kindred spirits in
their typological approach to physical anthropology, shared ideas
about human evolution, interest in manlike monsters, and a mutual
dislike of Washburn.[54]

In the summer of 1974 Krantz and Coon sat down for a long
conversation that ranged far and wide over their work, their colleagues, and the state of anthropology. Coon clearly liked Krantz
and saw him as an acolyte. Their relationship had become comfortable enough for Coon to let his guard down and share his deeper
feelings with the 43-year-old Krantz. He told Krantz he was grate-
ful for the plaster casts of Sasquatch prints he had sent him, including the Cripplefoot. He saw it as a "pleasure to discuss these things
with somebody who knows what he is talking about." Still smarting
deeply over the *Origins of the Races* controversy, Coon called Ashley
Montagu "a fake in about all ways." He referred to William C. Boyd
(1903–83) as a "lonely, poor, old man" whose books "should not
have been published." Boyd argued that blood types held crucial, if
not primary, information on the relationship of different human races,
something Coon and his mentor Hooton strongly disagreed with. He
considered William Howells "o.k.," but called Sherwood Washburn
"a slimy bastard" with "no original ideas," who had learned everything he knew from himself and Ernest Hooton at Harvard. Having
almost had his career destroyed by Washburn, Krantz did not disagree. Coon then questioned the intellectual capabilities of some,
and the sexual orientation of others, in the world of anthropology.
He recounted an odd story of how he once accidentally passed wind
in the face of Theodosius Dobzhansky at a dinner. Krantz opened
up, too. He told Coon about the women who became his first three
wives. When Coon asked who they were, Krantz responded grimly,
"nobody important."[55]

Coon and Krantz also talked of race in their tête-à-tête. After stating that Israelis could not measure up to the morality of Arabs, he
asked Krantz how his racial theories had been received. Coon insisted
his ideas put forward in the *Origin of the Races* did not support racism, nor was it meant to. Coon elaborated that he did not support

the UN resolution on the equality of races, though he did support the notion that individuals of any race could, and did, achieve intellectual heights as easily as any other. Coon could also commiserate with Krantz over the fact that other anthropologists rejected their monster work. In a tone that would have made Ivan Sanderson and Bernard Heuvelmans proud, Coon asked Krantz rhetorically "What could stuffy academic pedants do if they had to cope with new ideas every few minutes?"[56] The mainstream considered what both men did as existing in the murky and dubious realm of pseudoscience. In the latter twentieth century, race and monsters counted as antiquated nonsense from a bygone era.

It is difficult to discern for sure what place politics occupied in Krantz's world. In addition to his ambiguity over Carleton Coon's views, he is difficult to pin down on other issues. For example, as a military veteran, he opposed the war in Vietnam. He enthusiastically wrote a letter of support for a student who applied for conscientious objector status, but when arrested as part of an antiwar protest at Berkeley, Krantz vigorously argued to the judge that he was not a protestor but had simply been in the wrong place at the wrong time. Krantz made no overtly racist statements in any of his published work or correspondence and did not support any aspect of eugenics. He avoided the racism of segregationists, stuck to his own more moderate views, and negotiated his way through the battles over race engaged in by his mentors, Washburn and Coon, and continued to see the practical uses of typology for his monster work. His work on Sasquatch occupied the central focus of his life and any techniques or persons who supported that work, regardless of what baggage those techniques or people carried with them, he accepted. This single-mindedness he shared with his nemesis, René Dahinden, who also determined his relationships based upon whom or what aided his work. Krantz had little choice in this matter, however. Not a racist, Krantz needed the Carleton Coon style of racial origins and evolutionary *continuity theory* if he were to make sense of the origins of Sasquatch. Along with typology, in his own way Grover Krantz also brought the ideas of The New Physical Anthropology to monster studies. He adapted Sherwood Washburn's emphasis on anatomical function to the study of Bigfoot through his continuous attempts to work out the biomechanics of Bigfoot anatomy. Just as the New Physical Anthropologists saw human ancestors in terms of evolutionary adaptation to environment, Krantz did the same for manlike monsters.[57]

Convinced of the reality of Sasquatch, Krantz increased his efforts to gather more evidence to convince his colleagues. He collected plaster casts from other investigators and created the first life-size reconstruction of a *Gigantopithecus* skull and jaw for comparative and illustrative purposes.[58] He theorized on the life of an anomalous primate in notes labeled "growing up Sasquatch." He speculated about the size of family groups, rogue males, territoriality, eating habits, and other aspects of presumed Bigfoot behavior.[59] He worked out the creature's weight, its height-to-stride length ratios, and foot structure. Along with bones and a mock-up of the creature's form, Krantz knew he also needed a theory to explain the existence of Sasquatch. The crackpots could get away with just chasing around the woods. As a scientist he had to construct a theoretical evolutionary framework in which to situate Bigfoot in relation to other creatures in general and some species or genus in particular.

Positing *Gigantopithecus* as theoretical progenitor of Sasquatch—Krantz as well as others thought—would make sense of all the evidence. Krantz hypothesized that *Gigantopithecus*—or a *Gigantopithecus* variant—had managed to migrate across the Bering land bridge along with other Asian fauna to trickle down into North America. He had initially learned of *Gigantopithecus* as a graduate student when he read the work of Weidenreich and von Koenigswald. Once his interests turned to Sasquatch, Krantz read their work more closely, imbibing a good bit of their ideas on human evolution and modifying them along the way. From this point on *Gigantopithecus* dominated Krantz's thinking, but he had plenty of company.

While academics like Carleton Coon and George Agogino, and university trained investigators like John Green, knew of this creature, it is unlikely any of the North American Sasquatch hunters considered *Gigantopithecus* prior to reading Bernard Heuvelmans or John Green (a segment of that community never accepted the idea).[60] A number of anomalous primate enthusiast authors made brief mentions of the work of Weidenreich or von Koenigswald after Heuvelmans but did not pursue the issue.[61] These derivative works fall into two main types: the pseudo-scholarly and the sensational. Don Oakley and John Lane referenced *Gigantopithecus* in a Yakima, Washington, newspaper article in 1960.[62] Michael Grumley's *There Are Giants in the Earth* (1974) takes a less sensational tone as well. He discussed the *Gigantopithecus* theory and offered a unique alternative, suggesting the possibility that the creature had evolved in the Americas and used the Bering land bridge to migrate *into* Asia instead of out of it. "The only claim that may be made," he says, "is that it is possible that the ancient giant...began his wanderings in the Americas."[63] Grumley consulted a number of people for his book, including John

Napier and Grover Krantz. He gave Krantz credit for promoting the *Gigantopithecus* theory and for making "the present North American creature respectable" to study.[64] B. Ann Slate and Alan Berry make a blind reference to *Gigantopithecus*, saying "paleontology has provided evidence of likely ancestors for the giants witnesses now describe."[65] Their book, *Bigfoot* (1976), is more in the sensational range, as they also claim what makes the Sasquatch so hard to find is that there are "spiritual and other unknown forces at play" and that a UFO connection exists.[66]

## Krantz and Evolution

In pushing the *Gigantopithecus* theory furthest and most consistently, Krantz completely rejected the spiritual entity and alien visitor ideas, considering them the "lunatic fringe." This helps account for his attraction to the work of Heuvelmans, Sanderson, and Green, who searched for prosaic explanations. Increasingly curious about the phenomena, he began making visits to various places where the creatures had been seen. He was already well acquainted with *Gigantopithecus* from his university studies and reading Weidenreich and von Koenigswald, and with the Yeti connection from reading Heuvelmans and Sanderson. John Green's argument brought Krantz closer to the *Gigantopithecus* hypothesis, and the Bossburg evidence convinced him. The evidence built to a point of critical mass for him and he made the final intellectual leap. In a set of notes from the early 1970s Krantz explains his steps to conversion:

> Fall, 1964. First trip to Bluff Creek area No. Calif. Other trips over much of the next two years. Neanderthal theory prominent, but uncertain. Read Sanderson.
> Spring, 1967. Patterson's film still shot—uncertain. Still Neander.
> Spring, 1969. first jaunts from Pullman seeking evidence.
> Saw Patterson's film.
> Got Green's book. Gigantopithecus or Australo
> Summer, 1969. First interview with witness at Hoquiam.
> Winter, 1969-70 Colville [Bossburg] incident [67]

Though initially coming to the anomalous primate question because of his interest in Neanderthals, Krantz discounted the idea that Sasquatch might be a relic caveman. He explained his hesitation on the subject to Bernard Heuvelmans. The father of cryptozoology, who believed the Patterson film a hoax, hesitated in his original assessment of the film because of Krantz's analysis. After discussions with Krantz, the Frenchman wondered if he had not been wrong about the nature of the celebrated footage after all. "Nobody understands better than I," Heuvelmans told Krantz, "your reluctance to accept the Neanderthal hypothesis." Krantz's work also helped convince Heuvelmans that the *Gigantopithecus* theory he had originally assigned to the origins of the Yeti, held for Bigfoot as well. Nothing, however, could shake Heuvelmans's contention that as far as the Minnesota Iceman was concerned, "there cannot be the slightest doubt this is a Neanderthal Man of a very specialized type."[68]

Krantz began delving into Sasquatch studies at the same time as he learned the mechanics of evolution as a graduate student, and the two realms merged for him in subtle and not always compatible ways. Krantz articulated his own views on human evolution in his doctoral dissertation *The Origins of Man* (1971), and again in *The Process of Human Evolution* (1981), employing an idea reminiscent of one termed *regional continuity*.[69] This concept, also known as the *multiregional theory*, found its genesis in the work of Franz Weidenreich, but appeared in a more sophisticated form worked out in the early 1980s by evolutionary biologists Milford Wolpoff, Alan Thorne, and others. Wolpoff and Thorne held that instead of a single wave of already modern humans coming from Africa and pushing out archaic hominid groups that had gone before them—the *population replacement* theory championed by British paleoanthropologist Christopher Stringer—groups of *Homo erectus* left Africa and, in adapting to their local environmental conditions, exchanged useful genetic materials and evolutionary characters (genetic drift), and later evolved into modern humans in the areas associated with major modern ethnic groups.

While superficially similar to continuity theory, Krantz's work differed in subtle and important ways. For him the engine that drove *Homo erectus* to become *Homo sapiens* was the appearance of culture. The changeover began, he argued, when *erectus* populations developed tool making and in particular persistence hunting. Krantz saw *Homo erectus* as little different from *Homo sapiens* anatomically. He would later say, "No valid distinction between conventional *Homo erectus* and archaic *Homo sapiens*" could be found. The difference came in the possession of cultural behavior.[70] Krantz said an "uninterrupted gene flow" between *erectus* and *sapiens* existed. Using Weidenreich's term, but not necessarily his complete idea, he said, "ultimately it is a *continuum* [author's italics] subdivided only by the lives of the individual organisms."[71] Krantz also incorporated the emergence of language into his explanation of how *Homo erectus* turned into *Homo sapiens.* He claimed that as *erectus* individuals grew to maturity they acquired speech. This hastened along the acquisition of culture and thus evolutionary adaptation.[72]

Krantz built supportive theoretical structures for Bigfoot into *Climatic Races and Descent Groups* (1980), which ostensibly had nothing to do with anomalous primates. Krantz argued the origins of the races grew out of their adaptation to local environmental conditions. *Homo sapiens,* Krantz claimed, are too young for racial diversity to have appeared in their species's lifetime. The only alternative postulated modern racial diversity beginning in earlier *Homo erectus* populations. He asserted that modern ethnic traits "can already be seen in *erectus* times," arguing that populations of *erectus* having left their African home found their way to various corners of the world; then, because of the unique climatic conditions they encountered, they began to evolve into racial types as separate subspecies of *Homo sapiens.*[73]

From its introduction, multiregionalism ignited a vigorous debate within the world of paleoanthropology and became the counterpoint to population replacement. *Climatic Races* appeared in print a year before Wolpoff's and Thorne's first article on the subject, so Krantz could not have used it for reference. The structure of his theory came partly from the work of Weidenreich and more so from Carleton Coon. His short bibliography for *Climatic Races* reflects this, as it contains only the works of Coon. Afterwards, however, not referencing Wolpoff becomes more problematic. Whether or not Krantz's idea mirrored multiregionalism or just superficially resembled it, he never acknowledged the contributions of Wolpoff and Thorne.[74] By the mid 1980s it would have been difficult to do research into human origins without encountering this important debate. Krantz managed to ignore it despite the profound significance it had for his work on both humans and Sasquatch. As late as 2000 Krantz referred to Wolpoff as "Wollpot" in an interview, and only acknowledged him as "a very prominent fellow."[75] Employing a form of regional continuity that owed its genesis to Carleton Coon rather than Milford Wolpoff, Krantz worked out his explanation for the origin of the Sasquatch in an attempt to stake out his position as the foremost authority on anomalous primates.[76] This theoretical model allowed him to argue that an Asiatic *Gigantopithecus* population evolved into the Yeti, while another group crossed into North America and, in adapting to climatic conditions there, evolved into Sasquatch.[77] Krantz may have ignored multiregionalism because he felt it was unrelated to his work. As he never articulated a position relative to multiregionalism or population replacement it is unclear why he chose to ignore it. An answer to this question might be found in the fact that multiregionalism is largely genetics based, and Krantz never felt as comfortable with genetics as he did with traditional morphology. Milford Wolpoff also points out that Krantz followed Coon in describing a pattern of *independent* evolution of different races, whereas his position held that there were no races and that human populations from different regions evolved the same way because they exchanged genes. This is probably why he also did not cite Weidenreich.[78]

Complicating this aspect of Krantz's work, he and Milford Wolpoff knew each other as friends who spent time together. Krantz's wife at the time hailed from the Ann Arbor area home of the University of Michigan, where Wolpoff held a faculty position. When visiting his in-laws Krantz would visit Wolpoff. He gave Wolpoff plaster casts of Sasquatch prints and often discussed anomalous primate evidence. He did this often enough to make Krantz well known in the Michigan department to both faculty and students. Wolpoff liked Krantz and refused to ridicule his work the way others did. He also tells the story of how around 1981 a Chinese anthropology student, working on his English language skills, asked in class the meaning of the word "crazy." A discussion ensued, with other students giving explanations, often humorous, of the word and how to use it. One student then wrote on the board Grover Krantz's name and explained how, despite Dr. Krantz being quite sane, some considered his Sasquatch work "crazy." Wolpoff explained to the class how Dr. Krantz's work fell well within the bounds of what a scientist was supposed to do—explore the unknown—so that in this case crazy meant more an unorthodox pursuit rather than having a mental condition. The classroom conversation then moved on. As luck would have it, a few moments later Grover Krantz unexpectedly showed up on one of his visits. A student scrambled to stand in front of the blackboard to cover Krantz's name so that neither he nor the class would have to explain and be embarrassed.[79] Their personal relationship only makes it more difficult to explain Krantz's ignoring of Wolpoff's work.

Krantz saw the study of evolution as more than just looking at fossils on their own. In the introductory lecture to his class in human evolution at Washington State University during the 1990s, Krantz said "fossils are useless by themselves"; theoretical structures were crucial to their analysis.[80] He argued that the study of evolution contained cultural aspects. Human fossils, he said, had been found for years, but they only became "significant" when they fit into cultural as well as scientific frameworks. Physical evidence, he believed, said little of use without a corresponding paradigm. While discussing theories about how *Australopithecus* did or did not fit into human lineage, he chided his colleagues by saying that "emotionally based opinions...are translated into...how 'human' these Australopithecines really were."[81] He told his students "evolution satisfies a cultural need—origins," and so they would be studying "personal prejudices" as well as anatomy.[82]

At the conclusion of *The Process of Human Evolution*, Krantz discusses Sasquatch and lays his cards on the table. He makes it clear he is trying to construct an acceptable paradigm that would establish him as the primary authority on anomalous primates. "I have probably investigated the possibility of unknown hominoids," he boasted, "more thoroughly than anyone else currently active in the academic community." He continues: "Facts are interpretable only in the context of some preexisting theoretical framework."[83] Unfortunately, creating a framework that would make sense of the Sasquatch material proved no easy task. "It is understandably difficult for new concepts to make headway against this array of cultural and economic forces."[84] Then, playing the martyr, he concludes: "It has not been my purpose to write what people want to believe about human evolution, though that procedure often sells many books."[85] René Dahinden saw Krantz was jockeying for a position as academic leader. In an interview, Dahinden, always ready with a barb, said of his old acquaintance, "He certainly has a problem trying to be somebody."[86]

Though he did not believe Bigfoot to be a human ancestor, Krantz did believe that the study of the creature said something about human evolution. He felt so strongly about Bigfoot's nonhuman ancestor status that he disagreed with the man who saved his career, E. Adamson Hoebel, over the issue. In his *Man in the Primitive World* (1949), Hoebel argued that man's earliest ancestors, like *Meganthropus* and *Gigantopithecus*, were giants that evolved to progressively smaller size until *Homo erectus* reached modern human proportions. He wrote, "gigantism was subsequently progressively modified in a reversal of the usual evolutionary trend, which is from smaller to larger."[87] The suggestion is that humans descended from giant ancestors, the way Franz Weidenreich thought. In the margins of his copy of Hoebel's book next to this quote Krantz wrote "NO!"[88] Krantz argued that there were two schools of thought on how the human line relates to the rest of the natural world: either humans are just another animal or they are biologically unique and special. Scientists who tend towards the "just another animal" school believe humans are the product of rapid evolution from a recent ancestor, while the "unique" school sees slow evolutionary change from a very distant ancestor. Krantz argued that the former accept Sasquatch as a kind of link between the human line and nature, while the latter reject Sasquatch as a real animal because "they find it emotionally satisfying to believe that there is a vast unoccupied gulf separating us from the rest of the animal kingdom."[89]

## Conclusion

At the same time as his career took off, Krantz suffered one of the great personal losses of his life. In 1972 his beloved dog, Clyde, passed away. Their relationship had been so close he was inconsolable for days. Soon after the dog's death, Krantz and Evelyn broke up, and in 1975 he found himself divorced again.[90] Ever interested in bones, he wanted to have Clyde mounted, or at least have his skeleton recovered, but at first he hesitated doing it himself, unable to cut and rend the flesh of one so close to him.[91] Finally, he buried Clyde outside his home and allowed nature to take its course. A year or so later he dug out the skeleton and cleaned his old friend one last time. A few shovels into the exhumation the skull appeared. Sitting there alone in the dirt of his house's alleyway in the spreading gloom of the evening, Krantz contemplated the animal's skull, like Hamlet with poor Yorick, and philosophizing about the attachments we sometimes make to animals. Krantz, like Hamlet, pondered life's big questions as he held the dog's skull. "Maybe we shouldn't get so attached to other beings," he waxed poetic as he stared into Clyde's now empty eye sockets. In words that would be a prophetic description of his relationship with Sasquatch, Krantz wondered whether obsessions over such relationships could be harmful. He answered his own question with a statement that says much about his pursuit of anomalous primates in the face of ridicule. "If we didn't" have these relationships, he wrote, "we wouldn't really be human, would we?"[92] As a way of dealing with the loss of Clyde, Krantz wrote a description of their lives together. That, too, pained him and he did not finish it until 1981, then self-published it only in 1998—as he faced retirement and his own mortality—as a 32-page publication. *Only a Dog* is a sad love letter to his one best friend.[93] Over the years he would have a succession of wolfhounds, but he always thought of Clyde. In a way, Krantz substituted Sasquatch for Clyde, and lavished all his attention on the ape the way he did with the dog. It was unlikely that he would end up burying Bigfoot in his yard, so there would not be the same level of loss with the large furry two legged creature as he had with the smaller four legged one. And, he was not alone in his obsession.

# Chapter 5

# Suits and Ladders

*This, I must warn you, is a very hairy subject.*

Ivan Sanderson

In November of 1967 a graduate student assistant reached up to grab the cord on the window shade and pulled it down to block out the Manhattan daylight, tinged silver with the approach of winter. The conference room on the fifth floor of the American Museum of Natural History went dark. Most of the people sitting, waiting, in the room were scientists, researchers, and assistants from the departments of mammalogy and anthropology; the others included some wire service reporters. Some were impatient, some preoccupied with the work they had left in order to come to this meeting, others thinking of the rapidly approaching Thanksgiving holiday, others finishing what was left of their lunch. A man threaded a roll of 16mm film into the projector that sat on a table in the back of the room. The group had been rather awfully assembled to see this bit of film.

The man preparing the film had also shot it. Roger Patterson, a sometime cowboy, rancher, backwoods guide, film industry hopeful, and Sasquatch hunter from Yakima, Washington, finished preparing the machine and looked up expectantly at the group. According to Patterson, just weeks before, he and a rancher friend, Robert Gimlin, had ridden out looking for a Sasquatch in the thick forests of California, along a stream ironically called Bluff Creek, when they spotted the creature. Patterson grabbed his camera—after almost being thrown from and crushed by his horse when the animal reared in fright on encountering the thing—and began filming. Now Patterson, Gimlin, and Patterson's brother-in-law construction contractor Al DeAtley had brought the film to the great metropolis, to one of the most respected seats of scientific knowledge in the world, to show their film to the assembled experts.[1] The cowboys would genuflect before them and wait for the all knowing hands of science to wave the motion of validation over the film. The scientists were led by curator of mammals, Dr. Richard Van Gelder (1928–94), and Anthropology Department head, Dr. Harry Shapiro (1902–90). Van Gelder, a specialist in the evolution of mammals and curator of the museum for 25 years, also worked on endangered species. Shapiro, a specialist in human evolution, had designed the museum's Hall of the Biology of Man, which opened in 1961; he helped pioneer the field of forensic anthropology as well as the reconstruction of the dead in order to determine their identity. These men had the background to appreciate the film's significance and place a gold stamp of approval on it.

The cowboy from Yakima and his brother-in-law (Gimlin did not come to the museum) stepped into the hall while the film ran. The long hallway that connected the mammal and anthropology departments had both sides lined with polished wood and glass display cases and metal wall lockers, where the scientists kept their working collections. Average visitors rarely saw this space. It smelled of mothballs. As the door closed, a switch turned on and the rattle of the projector began. The light flickered briefly, and there on the screen the tableau unfolded. The image of an apelike creature, jumpy and fuzzy at first, evened out to show the thing dark and hairy walking upright, on two feet, away from the camera. At one point it turned to look directly at its pursuers, as if pondering these interlopers who had disturbed its bucolic reverie. Without breaking stride or hesitating for an instant, the thing strode purposefully off disappearing into the trees.[2]

It took just seconds. The rattling of the projector stopped, the flap-flap-flap of the film running out of the spool sounded, and the lights came back on. The room full of expert anthropologists, specialists in the workings of warm blooded animals, particularly man and his closest evolutionary relatives, the egghead's eggheads, said nothing. No one requested to see the film again. No rush of enthusiasm ensued, no frantic exchanges of comments, no popping of champagne corks; nothing. The group sat like a Greek chorus with no lines to read. Van Gelder asked the audience what they thought, and the briefest of discussions took place, the voices muffled and inaudible to Patterson and DeAtley waiting outside. After a few minutes the conference room door opened and Van Gelder and Shapiro stepped into the hall and politely thanked Patterson for bringing his film. Then the assembled audience shuffled out and went back to what they had been doing before lunch. Neither Van Gelder nor Shapiro, nor any of their staff, had seen anything noteworthy in the film. The general feeling among the scientists fell along the lines of the film being staged. The wire service people then stepped out, the looks on their faces even less encouraging. After a flurry of polite smiles, the two men from Washington found themselves in the hallway of the great museum all by themselves, holding a canister of film and nothing else. The sound of ambition faded with the footsteps of the retreating academics. Only the smell of mothballs remained behind. Patterson's grand hopes and expectations had been dashed.[3]

### The Cowboys

Roger Patterson (1933–72) and his somewhat reluctant partner, Robert Gimlin, had been monster hunters for some time before they headed to Bluff Creek. Born in South Dakota, Roger Patterson was soon transplanted when his family moved to Yakima, Washington. He played football for Yakima Valley College and then served with the U.S. Army in West Germany. Back in Washington following his military service, he married and began raising three children. Life in Yakima offered few opportunities for a man like Patterson, who had seen a bit of the world and yearned to be somebody. He always seemed to be up to one financial scheme or another.[4]

Patterson had dreams of entering the movie business or making it big somehow. Reading Ivan Sanderson's "The Strange Story of America's Abominable Snowman" in 1959 was a transformative experience.[5] He determined to go and find the creature. So taken with the idea was Patterson that he contacted Sanderson and they began a long correspondence that would benefit both in later years. Following the pattern for nascent anomalous primate enthusiasts, Patterson began scouring the media, collecting everything he could on big hairy monsters. In 1966, playing off Sanderson's article, he entered the fray by publishing the awkwardly titled *Do Abominable Snowmen of America Really Exist?* More a compilation rather than an original written work, Patterson's little book contained reprints of important newspaper accounts and articles written by others, with a bit of introductory commentary here and there along with a clutch of drawings Patterson had made himself. Indeed, it is this artwork that is the most original contribution Patterson made to the volume. He starts off the book with a quote of his own. Set apart from the rest of the text, it is reminiscent of the opening of *King Kong* which imparts an 'Arab Proverb' before beginning the tale. With the philosophical flourish of an intelligent man desperate to break free of his rural background, solemnly intoning, "He who seeketh long enough and hard enough will find the truth whatever that truth may be."[6] It is a line that could have served as Patterson's epitaph: or for any of the monster hunters.

In 1964 Patterson made his first trip to Bluff Creek. While there he met U.S. Forest Service worker Pat Graves, who told him he had seen many tracks in the course of his official duties in the area and took Patterson to view some. Excitement growing, Patterson immersed himself in Sasquatch studies and began to put his book together. While doing this he met John Green and René Dahinden in 1965.[7] After putting out his book the next year, Patterson planned to make a documentary film on Bigfoot. The choice of Bluff Creek as a place to do the filming came easy. Since Jerry Crew's famous 1959 discovery, there had been considerable activity in the area. In 1959 and again in 1960 another road builder, Bud Ryerson, came across Bigfoot tracks, with more found by others intermittently throughout the 1960s. In 1966, the same year Patterson published his book, a man claimed to have seen a Bigfoot crouched down and drinking from a stream, supposedly by using its hands as a cup as a human would. In early 1967 John Green and René Dahinden considered the activity near Bluff Creek strong enough to warrant a trip down from Canada. The tracks they found caught the attention of Don Abbott (1935–2005), an archaeologist with the University of British Columbia at Victoria. Intrigued by what he saw, Abbott managed to talk others at the university into meeting with Green and Dahinden, though it did not go well. Discussing the evidence laid before them, the academics tried to explain it away. This approach always infuriated Dahinden. Unable to retain his composure as the boffins discussed evolutionary mechanics, he burst out, "I'm sick of your scientific jargon!" Trying to salvage something out of the effort, Dahinden attempted to talk the UBC staff into hiring him and Green to search the area for more evidence. Exercising his deft gift for tact, Dahinden said, "You employ enough deadheads as it is; two more wouldn't sink the boat." Perhaps not surprisingly this approach did not sway the scientists and they parted company.[8]

About this time Bud Ryerson found tracks near Blue Creek Mountain just a short distance from Bluff Creek. Patterson and his friend, Bob Gimlin, in Washington searching for tracks around Mt. Saint Helens, in the area where an encounter between anomalous primates and armed men reportedly had occurred in 1924 (known as the Ape Canyon incident), took notice.[9] News of tracks at Blue Creek

Mountain brought the two cowboys back down from Washington to give Bluff Creek another go. Patterson thought he would take background film footage of the area for his planned documentary. They might even be lucky enough to film a footprint or two. Patterson rented a handheld 16mm movie camera from Sheppard's Drive-In Camera Shop in Yakima, grabbed Gimlin, three horses, supplies, and a truck and trailer to carry them all. They raised some operating capital from acquaintances, including Patterson's brother-in-law, Al DeAtley. In October they headed out to Bluff Creek. They pushed in as far as the truck and trailer would go, established a base camp, and then continued deeper into the wilderness, each man riding a horse with a third as a pack animal. Common images of California come from the urban sprawl of Los Angeles or San Francisco or the endless summer of the beaches. Northern California, however, is a region of rainforest density, with huge swaths of virtually impenetrable wilderness and little or no signs of civilization other than the occasional rough logging roads that wend their way through the rugged terrain (the type of road Jerry Crew and his fellows were building when they found the Bigfoot tracks). Part of the Six Rivers National Forest, the area of Bluff Creek lies in the North West corner of the state near the Oregon border. The cowboys had ridden a few miles further through this dense but beautiful terrain when they came around a bend in the trail that opened onto a clearing at the edge of Bluff Creek itself. As they approached the creek eternity suddenly stood up in front of them on two hairy legs and stared them in the eyes.[10] Recovering from his initial surprise, Patterson dismounted and managed to film the creature as it turned and strode away.

After the encounter, Patterson also filmed the tracks the creature had left, then made plaster casts of some of the better ones. Somehow, it is not clear how or by whom, the film went out for processing. Patterson then began to frantically contact everyone he could to tell them what had happened. They called Don Abbott at UBC and breathlessly instructed him to come at a run and to bring tracker dogs. Abbott, chastened by his bosses' reactions when he had brought John Green and René Dahinden to see them, declined, instead saying he would look at the film later. Word spread quickly through the Sasquatch enthusiast community that something extraordinary had happened. Green and Dahinden did come at a run. The film and its makers reached Yakima by October 22 and viewed it for the first time at DeAtley's home, along with Green, Dahinden and friend Jim McClarin. Gimlin did not attend this first viewing. Exhausted from the ordeal he went home to rest.[11]

The group watched the film over and over, but it was a bit of an anticlimax. René Dahinden saw in the film exactly what he expected to see, while John Green remembered that the film itself could not be described as impressive. The figure of Bigfoot seemed small, almost insignificant, in the wider film frame. Patterson wanted to head right away to New York or Chicago and show it off. Green and the others suggested starting out a bit more modestly. Dahinden prophetically said that if he ran off to New York he would be "laughed out of town," and seen as a "freak with a monster movie."[12] On October 26 Patterson screened the film for a group at the University of British Columbia. Despite their previous encounter with the UBC "deadheads" the monster hunters received a welcome back. The screening, however, produced the type of mixed reactions common among the scientific community. It did not bode well for the future.

## A Gamut of Orthodoxies

As he paced back and forth in his Manhattan hotel room Ivan Sanderson looked irked. His anger stemmed from the reaction Roger Patterson and his film had received at the hands of the New York scientists. Despite his own formal training and self-described status as a "zoologist" and "botanist," he regularly disparaged professional scientists and academics—calling them crackpots in *Abominable Snowmen*— for not enthusiastically jumping on the ABSM bandwagon the way he had. As is common amongst researchers in the dubious realms, Sanderson saw himself as one apart from the mainstream. He reveled in his status as an independent thinker not tied down by corporate structures or by "the whole gamut of orthodoxies" that arose when new discoveries came to light.[13] Patterson and DeAtley were deeply disappointed by the New York reaction, but Sanderson saw the incident as a sudden and unexpected opportunity for him personally. It created an opening he took full advantage of. He firmly believed the thing in the film to be real. Always ready with a jab at naysayers, he said, "I think it is genuine; *and so do most really competent anthropologists* [author's italics]."[14]

The reaction of the scientists at the American Museum of Natural History that November doubly disappointed Patterson and his associates. The reason for bringing the film all the way to New York was to sell it. As it had done with the Yeti, *Life* magazine showed interest in Bigfoot. It would run a cover story on the film and the men who shot it and, of course, pay handsomely for the right to do so. As a show of its serious intent, the magazine paid the expenses of the three men from Washington to come to New York. Before any real money would change hands, however, *Life* editors insisted that the film be given the stamp of approval by reputable scientists and preferably some institution of world renown. Few institutions of science had the clout of the mighty American Museum of Natural History. Conveniently, the museum rose imposingly (as if a Mayan temple had been built by Victorians) over several large city blocks just up the street from the *Life* offices. Telephones rang, markers were cashed in, cajoling engaged, and a private screening for the people at the museum hastily coalesced. The cool reaction of the museum staff put an end to it all. Scientists of the stature of Richard Van Gelder and Harry Shapiro did not risk their reputations, or the museum's, by putting their imprimatur on such a film only to have it found out to be a hoax later.

Disappointed by the reaction of the museum staff, *Life* hurriedly reached out to the New York Zoological Society (the Bronx Zoo). *Life*'s nature editor, Patricia Hunt (1922–83), corresponded on a regular basis with the zoo's curator of mammals, Joe Davis, over various animal issues, so the film was sent for a screening —without Patterson or his associates. Just as at the museum, it all happened very quickly. Davis received a phone call and shortly after a courier arrived with the film. He rounded up as many of his colleagues at the zoo as he could and they watched Patterson's footage. The animal behaviorists at the zoo, a staff whose training and daily work had them in close proximity to animals, including primates, saw nothing other than a cleverly concocted hoax. Years later, Davis remembered seeing what looked like the bulge of a zipper in the creature's back.[15] With this second poor showing, not only did *Life* back out of the deal, but its chief rival, *Look* magazine, which had been waiting hungrily in the wings, lost its appetite as well.

Almost since the day they had made the footage, Patterson and Gimlin wanted experts to see it. After the University of British Columbia screening, Patterson and DeAtley took it to Universal Studios in Hollywood to have special effects people study it. Then they showed it to a number of individual scientists and wildlife experts around British Columbia and California. By the time they reached New York they had experienced a very busy few weeks, indeed.[16] There, they did not get the reaction they hoped for. They really needed a cheerleader; they needed someone like Ivan Sanderson.

Jim McClarin, present at the first screening of the Patterson film in DeAtley's basement, alerted Ivan Sanderson to the film's existence and its imminent arrival in New York. Sanderson immediately called his contacts at *Look* magazine and the swarming began. When

Patterson arrived in New York he contacted Sanderson. The echo of the call had barely receded when Sanderson headed from his New Jersey enclave to Manhattan. The little group got themselves a few rooms then bounced around Manhattan for a few days trying to drum up excitement for the film. With *Life* and *Look* interested, it seemed like a bidding war might break out. Sanderson later insisted that Patterson and his companions wanted only enough money to cover their expenses and to mount a proper expedition back to Bluff Creek.[17] An article published in London in November did not mention the American Museum screening, but claimed that Patterson and the others "have already sold the television rights to their pictures for $50,000 (£20,800)."[18] With rejections by both the American Museum of Natural History and the Bronx Zoo, all that seemed dead in the water.

Seeing his chance, Sanderson stepped forward and took action. He contacted his editor at *Argosy*, Milt Machlin, and explained the situation. Within a few hours *Argosy* publisher Hal Steeger handed Patterson and Gimlin a contract, which they signed. The first big full color showing of Patterson's and Gimlin's Bigfoot would be on the cover of *Argosy* magazine. Sanderson, while upset at the reactions of the scientists, worked them to his advantage. If *Argosy* ran the story, Sanderson would write it. He worked quickly, turning a finished draft of the article in to *Argosy* by early December. Originally titled simply "Abominable Snowman!" the magazine ran it as the somewhat clunkier, "First Photos of Bigfoot: California's 'Abominable Snowman.'" It carried a cover date of February, 1968, but actually appeared for sale in January. Sanderson received $750 for the article.[19]

The deal with *Argosy* allowed Sanderson not only to write the story of the Patterson film, but to also take credit for having saved it from "the wipe." American anomalist author Charles Fort (1874–1932) did pioneering work regarding the more outré side of human knowledge and its relationship to the mainstream. He argued that strange discoveries challenged preconceived notions of science. When such material came to light, Fort said, the mainstream worked to eliminate, or wipe, it from view rather than have to answer uncomfortable questions about it. Sanderson could now crow that he had saved the Patterson film from such an ignominious fate.[20]

Joe Davis, of the Bronx Zoo, was not surprised Sanderson had become involved with the Patterson film. The two men had met in the early 1950s as members of a cave exploring club that held its meetings at the American Museum of Natural History. Davis liked Sanderson as a person quite a bit, but found his work problematic.

He complained—as many did—that Sanderson's scientific rigor fell somewhat short of the mark and that he tended to make glib statements about animal behavior, statements not always backed up with evidence. While Sanderson longed for the position and respect of a scientist, the work-a-day requirements of such a position did not interest him as much. His life hinged on the *role* of scientist rather than the job. He did not check his facts and this led to discrepancies that were often passed on by those who read and referenced his books. He seemed more concerned with his prose than his science. Running as fast as he could to keep up, he often seemed to be simply wrong on some of his zoological assertions. Davis insists that Sanderson never intentionally deceived, but often went off half cocked. Sanderson's reputation suffered less from others resenting him than from his own sloppy work.[21] More care taken with his basic science would have seen his ideas better received and his career less belittled.

Despite these shortcomings, Ivan Sanderson established himself as the first effective promoter of the Patterson film. Sanderson, argues Joshua Buhs, "made the beast into a consumer object."[22] Like many, he saw the potential for the film to be a goose laying golden eggs. His life may have been one long exciting adventure, but Sanderson never managed to convert it into monetary wealth. Always strapped for cash, he moved precariously from one writing job to another. He developed an ingeniously sly *modus operandi* for getting work. He parleyed his celebrity and knowledge of a field (like zoology, monsters, UFOs, or other odd phenomena) into getting work putting out articles for popular magazines. His breathless, over-the-top prose would draw attention and boost readership. He would then ingratiate himself with the publishers, and if an offer as an editor did not materialize on its own, he would suggest it. He then had a stable position for a while. His career being completely supported by his writing, he engaged literary agent Oliver Swan of New York's Paul Reynold's Agency.[23]

After they published Sanderson's *Abominable Snowmen*, Chilton Publishing hired him as an editor. Based in Philadelphia, Chilton began in the 1920s as an automotive repair periodical. As it grew, the company branched out into nonautomotive publishing with a catalog of quirky sensibility. In addition to Sanderson, the company published science fiction author Frank Herbert's first book, *Dune*. In 2005 Chilton released an edition of the *New Testament* in the original Greek. The automotive field was always the company's roots and primary source of income, and Chilton is famous for its extensive list of repair manuals geared towards self-reliant car owners. Some of the company's decisions frustrated Sanderson as an editor with ideas of his own. The initially good relationship soured, and he left.

Unable to acquire a university or museum position, Sanderson constantly hustled for work as a writer. He told friend Hobart Van Deusen of the American Museum of Natural History, "You have a steady job; I do not; so I must scramble to keep ahead of the mere process of living."[24] In 1949, when his popular radio and television shows went off the air, Sanderson provided a form letter to listeners showing them how to contact the networks to get him put back on, and it worked.[25] Following the Patterson film piece for *Argosy*, he wrote numerous articles on anomalous animals and phenomena for the magazine, earning between $500 and $1,500 each.[26] He became *Argosy's* science editor in 1969 and, in addition to feature articles, wrote the "Science Column" as well. Thanks to authors like Sanderson a steady stream of monster-related articles appeared in the many men's adventure oriented magazines of the mid-twentieth century. This publishing activity fed and encouraged a growing appetite for monster-related information.

A veteran at promoting his own career, Sanderson applied those talents to the Patterson film. Not long after the release of the *Argosy* article early in 1968, the popular "Joey Bishop Show" and the "Tonight Show" aired the film for American television audiences.[27] Sanderson screened the film at Emory University's prestigious Yerkes Primate Center in Atlanta (where William Charles Osman-Hill had taken up residence), as well as at the Smithsonian Institution, where John Napier (then head of the Primate Biology Division) saw it for the first time along with Russian anthropologist Vladimir Markotic. In July a documentary appeared on BBC television with Napier as host. Sanderson managed to get the BBC interested in the footage by going through filmmaker Richard Attenborough. A camera crew came to Sanderson's New Jersey farm to film him and John Napier as they discussed footprint casts. The film aired on the evening of July 27th.[28] Sanderson was tireless in lining up influential audiences to view Patterson's brief footage. The effort brought more sympathetic eyes in the scientific mainstream to the film, but with few exceptions even this generated little more than mild interest. These specially picked viewers, most trained, experienced, respected anthropologists and primatologists, went largely unconvinced.

Despite his ranting about the perfidy and cowardice of scientists at the American Museum over their opinions on the Patterson film, in private Sanderson had an otherwise generally good relationship with the museum. As mentioned, he counted as a friend museum mammalogist and ornithologist Hobart Van Deusen (1910–76) who in turn carried a membership card for Sanderson's Society for the Investigation of the Unexplained.[29] Sanderson invited Van Deusen to the dinner party welcoming Bernard Heuvelmans to the United States in October of 1968. Van Deusen likely took part in the screening of the Patterson film at the museum. That did not stop Sanderson from remaining friendly with him. The American Museum itself went on to have a somewhat awkward relationship with monsters.

## The American Museum

In the audience that day in November 1967 at the American Museum of Natural History viewing of the Patterson film was mammal evolution and behavior specialist Dr. Sydney Anderson. He later commented that in viewing the film, he noted the "proportions of the limbs and other parts of the animal…were quite compatible with those of a human being."[30] A few years later Richard Van Gelder stepped down from the chair of mammalogy and Anderson replaced him. While no public discussion of the film's showing at the museum appeared at the time (not even by the journalists present), word eventually leaked out, and in 1974 a *New York Times* reporter called on Anderson. The *Times* article that resulted discussed a number of Sasquatch-related issues. It mentioned Grover Krantz and Roderick Sprague as well as the work of John Napier. It also quoted Anderson as saying the creature in the film "looked like a man dressed up in a monkey suit," which is exactly what Grover Krantz had said the first time he saw the *Argosy* photos.[31] Anderson may have later regretted getting involved because his name had now appeared in print as a scientist with opinions on Bigfoot. This made him and the museum targets for enthusiastic monster hunters, and it fell to Anderson to deal with the various correspondence that came in.

A spate of requests came in the mid-1970s. A typical example originated in North Miami, asking for financial support for a proposed monster hunt in Florida. Anderson replied that "the department of mammalogy of the American Museum of Natural History [is not] in a position to finance the work you describe in your letter."[32] One from Texas asked the museum to sponsor a hunt in the Rocky Mountains, aiming to "track, film and possibly sedate "Bigfoot" long enough to bring him back for further study."[33] Anderson made the same reply to both. He quickly grew fed up with all the talk of hairy ape-men and requests for museum money and sponsorship to chase them. In March of 1977, Susan Hassler of *Quest/77* magazine wrote to Anderson. Her research for an article had led her to contact him. She asked him to "write a few words outlining your opinions as to whether or not these creatures could or do exist." Anderson gave her exactly what she asked for. Scrawled tersely across the bottom of her letter he wrote (so that his secretary could type the reply) this brief response, "I don't think they exist. My fee for writing is 10 cents per word."[34]

At this point Sydney Anderson had heard all he ever wanted to hear about Bigfoot, but he had another monster to deal with. The tabloid newspaper *Weekly World News* ran an article in April of 1980 titled "Bigfoot Photo Baffles Experts!"[35] It wasn't clear just which "experts" had been baffled, but the article included a photograph of a decomposing, severed head. The article told the story of the residents of Lewiston, New York (near the Canadian border), who in 1978 reportedly saw something beastly and strange skulking about the woods near their town. Described as a bear-apelike creature, the thing excited much interest and comment. A local police officer spent some time unsuccessfully stalking through the woods searching for it. In September of that year some hunters found the thing dead. They wanted to bring it back but none wanted to have to carry the rotting carcass home. With on-the-spot expediency they hacked off the head and left the rest behind. They took photos of it and wrenched out a few of its teeth as souvenirs. Having second thoughts about their prize on the trek home, one of the hunters gave it to a local boy who wedged the head in the lower branches of a tree in his backyard to allow nature to rid it of its putrid flesh. When he returned to check the head's progress, the boy found that raccoons had made of with the thing in the middle of the night never to be seen again.[36]

Eventually, news of the discovery made its way to Sasquatch investigator Jon-Erik Beckjord. He managed to obtain the photos, telling the *Weekly World News* that "We are very excited by the find." He then sent the photos to Sydney Anderson. Looking at the photos, Anderson sighed, rolled his eyes, and tossed the pictures on his desk. Anderson then received a letter from another Sasquatch enthusiast, Paul Bartholomew, who had heard of the upcoming article about the Lewiston head. Bartholomew's business card announced him as an "investigative researcher of unexplained phenomena." He told Anderson : "I have started a personal (funded by no group) investigation of the [Lewiston] incident." He asked excitedly "can it be the legendary BIGFOOT type creature?"[37] Anderson replied, "What I saw was a photograph of the severed head of a black bear (*Ursus americanus*), nothing strange about that."[38]

Several scientists made the list of those to contact regarding monsters, not just those at the American Museum. John Napier, too, received requests to verify and look at images. While his work took him some distance from anomalous primates, he retained an interest in them. In the mid-1980s *BBC Wildlife Magazine* contacted him over a story and some photos they had received. An Englishman named Tony Wooldridge had contacted the magazine with a startling story and photos. He claimed he had just returned from Northern India, where he had been involved in charity work in the Garhwal Himalayas and had encountered a Yeti. Outside the village of Gangaria in 1986, at 11,000 feet, he saw a creature about two meters tall, covered in hair, standing on two legs, skulking behind some vegetation. Wooldridge observed the creature for some time and approached it to a point about 150 meters distant. He also managed to snap two photos before he moved off, leaving the creature still in the same spot.[39]

The BBC magazine editor asked Napier to examine the statement and photos and provide a comment. (Wooldridge also sent materials directly to Napier). In a rough manuscript version of his reply, Napier said that after examining the material "my conclusions are remarkable, but quite logical." In *Bigfoot: The Yeti and Sasquatch in Myth and Reality*, Napier wavered in his opinion of the creature's reality. Read one way, he believes, read another, and he does not. Now, however, no doubt remained. He ruled out the thing in the photo being a bear or a human. That done, he felt he had no choice but to say, "After many years of doubt and partial disbelief I am now a Yeti-devotee." What else could the creature in Wooldridge's photos be? Napier fell into the same trap he had faulted Bernard Heuvelmans and Ivan Sanderson for falling into a few years before over the Minnesota Iceman. His enthusiasm got the better of his considerable observational skills. Some years later, a retracing of Wooldridge's trail found the creature still in the same spot. The Yeti in the photo turned out to be a snow-covered boulder.[40]

## Calculating a Monster

Neither filmmakers nor naturalists of any kind, Patterson and Gimlin failed to take down pertinent information about what had happened to them. In his excitement, Patterson even failed to record what speed he had his camera set at. This made it difficult to work out the creature's size, its speed, and other biological data that would have been of great help in any analysis of the film. As a result, it became something of a parlor game for various individuals to try to recreate the film by going to the site and measuring it from various angles. Taxidermist Bob Titmus arrived on the scene first. The man who had years before given Jerry Crew instructions for how to cast the first Bigfoot tracks, Titmus cast many of the Bluff Creek tracks still extant. John Green also did an early assessment of the site.[41]

The scientist who expended the greatest effort to analyze the Patterson Film was Grover Krantz. He never tried to debunk the film itself one way or another. After his initial disparaging of the footage he came to believe it genuine. As a result, he used the film as an artifact to help explain the biomechanics of the creature in the frames. He took the authenticity of the film itself as a given. This approach to Sasquatch evidence always proved problematic for Krantz. His penchant for taking evidence at face value haunted him all his career. He acknowledged the existence of faked evidence and misinterpretations of other phenomena, but once he accepted something as genuine he moved on and no longer questioned its validity the way a zoologist or animal behaviorist normally does not question the film of a subject animal in the wild. Their focus is in the behavior the film shows, not the film itself.

As an anatomist Grover Krantz saw that one way to lay the theoretical groundwork for Bigfoot involved showing it followed proper biomechanical principles. This would show it did not move like a man in a suit. Despite Bigfoot being a biped, if a real organism, it would be built and move differently than a human. By using the footprints and the walking gait seen in the film he could show the evidence made sense of the eyewitness accounts. To this end he did a series of mathematical calculations to determine not so much that Sasquatch did exist, but that it *could* exist. With comparative anatomy techniques used to reconstruct fossil animals, he made meticulous measurements and examinations of the footprints to see if the types of prints associated with Sasquatch fell within the parameters of a real animal. "An 8ft. tall, heavily built hominid," he stated, "would require certain structural modifications to its feet because of its great absolute body weight."[42] In other words, if evolution created a Bigfoot-like creature, what foot morphology would it have? He proceeded to work out ankle-to-foot-length ratios and other elements of bipedal walking biomechanics. As a baseline, he used physical data from "stoutly built men" taken from controversial "posture photographer," psychologist, eugenicist, and somatotyping researcher William Sheldon's *Atlas of Men* (1954) in order to build up a reasonable outline of the creature's curves. Krantz stated: "As body weight increases with the cube of linear dimension, cross-sectional areas in the limbs must also increase with the cube of linear dimension." He determined that in the Patterson film two frames gave a view of the creature's feet. He saw the foot's length extended twice the ankle diameter. This ratio, according to Krantz, fell just where biomechanical theory said it should. [43]

How would a backwoods bumpkin hoaxer know to do all this? A lucky guess might account for getting some things correct, but not this level of anatomical sophistication, Krantz argued. His notes are full of obsessive calculations and drawings showing him trying to work out the physical dimensions of the Sasquatch. Patterson used a Kodak K100 camera, but no one knew what speed he set it at. Krantz generated more reams of calculations at various film speeds to work out the creature's size and walking speed—he thought Patterson must have been shooting at 18 frames per second. His calculations also took into account possible heights and weights of the animal. By the early 1980s he determined to his satisfaction that "Patty," as the subject had been dubbed, was 6 ft. 8 in. tall, and that "there is no question that Patty is less than" 7 ft. 4 in. tall, and more than 4 ft. 2 in.[44] Krantz compared his results to the work of others, particularly D. W. Grieve of the Royal Free Hospital School of Medicine, London, and Dmitri Donskoy of the Institute of Physical Culture in Moscow. Both did extensive calculations and came to the conclusion that the film showed a genuine creature, not a man in a suit.[45]

Krantz managed to get a complete copy of the film for his own study as well as a used, crank operated film-editing machine so he could study each frame in detail. He reconstructed the path Patterson took at the site as well as the path the creature took moving away from him. He put together a list of all nine hundred plus frames with notes as to each one's clarity. He intended this work to be the heart of his treatise, *Big Footnotes: a Scientific Inquiry into the Reality of Sasquatch* (1992).[46]

To support his calculations Krantz gathered as much material as he could find on the camera Patterson used. Peter Byrne told Krantz that in a conversation with Patterson's wife in 1993 she said that in 1967 the only camera she ever saw her husband handle was the one he used to shoot the Bigfoot film. Specifically, he rented the Kodak Cine K100 (serial #3645) from a shop in Yakima on May 13, 1967. A common and popular 16mm windup model, it came with three interchangeable lenses mounted in a forward turret. The manual lists its film speeds as 16, 24, 32, 48, and 64 frames per second (FPS). Byrne confirmed this through official police records. Shortly after Patterson made his Bigfoot film, the Yakima police arrested him on suspicion of robbery. He had failed to return the rented camera, and the shop owner complained. Red faces appeared all around. Patterson immediately returned the camera with profuse apologies and the authorities dropped the charges, but not before a police report inventoried the camera equipment. Byrne excitedly told Krantz of his discovery. Krantz also found a copy of the camera's manual. It states, "For normal screen action, when using a silent projector, use the 16 frames per second speed."[47] Byrne, too, thought the film ran at 18 fps.[48] Oddly, both Krantz and Byrne concluded Patterson had shot at 18 fps. The manual, which both researchers had access to, does not list 18 fps as a speed option for that type of camera.

As usual in the world of Bigfoot studies, Byrne cautioned Krantz about his discovery of the camera details. Byrne happily shared with him, but "I see no point [at this time] in my releasing this information to others."[49] Byrne had long experience with the eccentricities and infighting that regularly took place in anomalous primate studies and wanted to pass this sage wisdom along to Krantz, whom he genuinely liked.

In his report on the Patterson film, Krantz took issue with the notion that what looked like breasts on the creature—which accounts for its commonly being referred to as Patty—might actually be a baby clinging to the creature's chest. This claim originated with researcher Jon-Erik Beckjord (1939–2008), a man whose cantankerousness and feuding with others in the field rivaled René Dahinden's, only more so as Beckjord later made extensive use of the Internet to attack his foes.[50] Beckjord also positioned himself as a vocal member of the school of thought arguing for the status of spectral entity and "shape shifter" for Bigfoot. He often told audiences at his talks that the jumpy nature seen in blow ups of the film came not as a result of the low resolution of the film stock, but as the record of the creature changing back and forth from one form to another in order to elude Patterson. Loren Coleman and his coauthor, Patrick Huyghe, have pointed out that "Beckjord has been arrested at, banned from, and thrown out of almost every serious scientific Sasquatch and cryptozoology meeting he has attended."[51] Krantz rejected the notion of the astral being status for Sasquatch. Wanting to stay as far from what he called the "lunatic fringe" as possible, he preferred prosaic explanations and referred to Beckjord as "an erratic individual who ought to be ignored."[52]

## Who Owns Patty?

The Patterson film had two intrinsic problems, the first is the question of its authenticity, the second its ownership. If genuine the film held the potential for generating a considerable amount of income through publishing royalties. Krantz became friends with Patterson, who confided in him sheepishly that he had sold rights to the film a bit haphazardly. Patterson did not possess formidable business acumen. He tended to go for the short term profit rather than a carefully worked out plan to take advantage of the treasure he had acquired. A number of people bought rights, unaware they conflicted with rights sold to yet others. As the man who shot the film, Roger Patterson owned the rights to it and so could do whatever he wanted. However, his companion on the trip, Robert Gimlin, found himself in a strange position. He felt he deserved a full share of the rewards. Like two people who go into a store to buy what turns out to be a winning a lottery ticket, one argues that he actually purchased the ticket and that the other simply came along. Holding the short end of the stick, and despite his friendship with Patterson, Gimlin found himself eventually bumped out like a superfluous acquaintance beyond his usefulness. John Green asserted: "As a result of a disagreement Bob Gimlin was cut out for several years from any participation in the profits."[53] Patterson and DeAtley had formed a company, Bigfoot Enterprises, without Gimlin's participation before they went to New York in November 1967. In response Gimlin, a little bewildered by the treatment, but willing to give his friends the benefit of the doubt, sued DeAtley and Patterson's widow in 1975 (Patterson had died in 1972), claiming he had been shorted on his portions of the profits from the film (the case settled out of court in 1976).[54] By June of 1968 John Green and René Dahinden had arranged with Patterson to get the specialized rights to lecture on the film in Canada.[55]

Grover Krantz, too, wanted to ride the Patty gravy train, at least for a while. He considered putting together a mail order business to sell plaster Bigfoot casts and stills from the Patterson film. When the U.S. Attorney General's Office looked into the matter of copyright of the Patterson film by reviewing the 1976 ruling, they found that Gimlin and his wife had been awarded 51 percent of ownership of the film as well as "100% of all publication rights of Bigfoot materials." Patricia Patterson held 49 percent ownership and 100 percent of television rights, while René Dahinden's position was rather weak and could be challenged. When Patterson died the rights to the film itself passed to his widow, Patricia. When Dahinden passed away the rights he claimed went to his estate.[56]

## Deconstructing Patty

The monetary value of the Patterson film, like any artifact, hinges on its authenticity. In 2004 writer Greg Long released the controversial *The Making of Bigfoot: the inside story*. Long set his goal as clearing up the murky nature of the Patterson film's origins. For years questions had surfaced, not only about the speed of Patterson's camera, but about the film's processing history, how it proceeded so quickly and mysteriously from field to lab, and what happened to other film shot that day? What course did the film take in the few days from the moment Patterson removed the roll from his camera to the moment it was shown in Al DeAtley's basement? The discrepancies and holes in Patterson's telling left even hard core believers wondering. By the time Long finished his version of events he had thoroughly outraged most of the manlike monster community. He interviewed the still-living primary characters of the story as well as a number of actors who had never been credited before. Long's contentious conclusion sketched Roger Patterson as a huckster, a cheat, and a con man, and his celebrated film as a hoax, nothing more than a man in a monkey suit. To prove his point, Long brought forward not only the man who supplied the suit, but the man who wore it.[57]

Long claimed Bob Heironimus, who had known Patterson for years and had been promised a thousand dollars for his part in the stunt wore the suit. Heironimus, a Yakima local had gone to high school with Patterson, and often to the local bar. He claimed everyone in town considered Patterson a deadbeat, and his film trash. The suit itself began as a theatrical gorilla costume supplied and customized by magician and TV horror personality Phillip Morris. Supporting the book, Long's associate and fellow phenomenology exposé writer and debunker, Kal Korff, wrote an article in *Fortean Times* defending their position and detailing how he, Long, and a camera crew from *National Geographic* television re-created the film by having Bob Heironimus wear a suit constructed by Philip Morris. Long's book and the re-creation constituted, Korff argued, "a breakthrough that cannot be dismissed, as it finally solves the case."[58]

The monster enthusiast community reacted with derision and disdain to Long's book. They quickly circled the wagons and began picking holes and pulling out inconsistencies in Long's narrative. They questioned times, dates, names, and recollections Long referenced. They argued that some of the people Long interviewed fabricated details, confabulated events, and generally played Long as a rube. Leading the charge in the same issue of *Fortean Times* as Korff's article, Daniel Perez wrote a scathing review of the book and of the re-creation. He kept up a steady barrage of insults in his own *Bigfoot Times* newsletter. In the April 2004 edition of his newsletter, Perez interviewed Peter Byrne, who said Long's book—in which he appears—"was amateurish, impetuous, and careless." Motivation for writing the book, Perez claimed, came from simple greed (the same argument for why Patterson made the film in the first place).

Perez pointed out that the television documentary "World's Greatest Hoaxes," produced in 1998 by Korff associate Robert Kiviat, used a sequence in which Clyde Reinke, a rancher turned documentarian, states that insurance salesman Jerry Romney wore the suit. Kiviat wrote a glowing jacket blurb for Long's book. This suggests they could not decide who wore the suit, Heironimus or Romney. Perez also claimed the suit used in the *National Geographic* re-creation, supplied by Morris as a duplicate of the original, looked nothing like the "suit" of the Patterson Film. Both *Skeptic* magazine and *Skeptical Inquirer*, longtime foes of anomalous primate studies, also pointed out inconsistencies and problems in Long's work.[59] The entire episode was a collection of charges, counter charges, and gainsaying by the various sides. In the film's defense, Long builds his reasonable-sounding argument on hearsay evidence and ancient recollections. While the parts are intriguing, he had few actual bits of incontrovertible evidence linking them together: no receipts for the costume, no test footage, no smoking gun or fingerprints. He had the same quality of evidence against the film's authenticity as supporters had for it.

The most ironic aspect of the reception of Greg Long's book, something missed by most of the individuals involved as tempers flared, took the form of a linguistic and philosophical turnaround. Bigfoot enthusiasts have always fallen back on the argument of authority. They enjoyed using the term "expert" derisively (not unlike the way creationists use the term "Darwinist" as code for incompetent scientist and cultural enemy). Scientists, as self-styled "experts" and "authorities," made easy targets. Their pontification, the amateurs argued, showed they knew little about Bigfoot. This way the amateurs could situate themselves as an embattled minority fighting the good fight against the collective power of the inflexible majority who stood ignorant of the facts and who would not listen to counter arguments. It gave the amateurs a sense of identity and mission. In describing the reaction of Bigfoot enthusiasts to Long's book and the

*National Geographic* re-creation, however, Korff turned the tables by making the enthusiasts the "experts." He summed up by saying, "We knew, going into our re-creation, that most of the supposed "experts" who still claim the Patterson-Gimlin film is real and have attacked our work, have never bothered to try and recreate the hoax for themselves."[60] After years of accusing scientists of not keeping an open mind about anomalous primates, Bigfoot enthusiasts like Daniel Perez and others were being labeled as the close-minded "experts" unwilling to listen to, and attacking, new evidence that did not support their side.

### Peter Byrne's Forgotten Analysis

In 1979 journalist Kay Bartlett managed to interview American Museum scientist Sydney Anderson about the Patterson film, despite Anderson's reticence to do so. When asked his opinion on the creature, he said that the Patterson film "was show business, not science. It was a man in a monkey suit."[61] Longtime veteran monster hunter Peter Byrne saw this quote and contacted Anderson, seeing in the scientist a kindred soul of sorts. Byrne had championed the film and the man who made it. Byrne, one of only a very few people who had direct connections to both the Asian and North American anomalous primate searches, and one of an even fewer number to actually search for the creature in both regions, held a special place of respect among many younger monster hunters. Following the expeditions for Tom Slick in Nepal, Byrne came to the U.S. to lead Slick's Pacific Northwest Expedition. Being an outsider brought in to tell the locals what to do did not always ingratiate Byrne to the older generation of North American enthusiasts like John Green and René Dahinden.[62] When the expedition ended, Byrne headed back to Nepal to pick up his game hunting and trail guide career. He came back to the U.S. briefly in 1968, but soon returned to Nepal, setting up the International Wildlife Conservation Society, Inc. That done, he returned permanently to the U.S. in 1970 and by 1972 had moved to The Dalles, Oregon, and resumed his hunt for hairy humanoids.[63]

In *The Search for Bigfoot* (1975) Byrne gave his assessment of the Patterson film. He said he had met with both Patterson and Gimlin and talked extensively about the circumstance of the film. He viewed the film numerous times, went to the site at Bluff Creek, and rued the fact that "not a single scientist in this country has taken the time to inspect the film."[64] He did give credit to Hobart Van Deusen of the American Museum for at least not dismissing the film outright, and to John Napier for his work. He came to the conclusion that Patterson and Gimlin did not have it in them to fake the film, and that the spot on Bluff Creek where the encounter took place could not have been a worse spot to stage a hoax due to its relative openness and regular use by rangers, hunters, and hikers. He said "There is no doubt in my mind...that the 1967 footage is a vital link in the chain of evidence, both hard and soft, that supports the existence of Bigfoot." He gave the footage a 95 percent chance of being genuine and the same percentage of the creature being real.[65]

Byrne's optimism about the Patterson film and Patterson took a turn for the worse in the few years after his book appeared. When he contacted Sydney Anderson at the American Museum in 1979, he had some dark things to impart to the scientist. Byrne told Anderson that his Bigfoot Information Center & Exhibit at Hood River, Oregon had done a further analysis of the film and now found it lacking. Anderson told Byrne that "as far as I'm concerned, all the other presumed evidence of such creatures is dubious at best."[66] The next month, Byrne publicly announced his retirement from the hunt. He had spent a lot of his own, as well as other people's, money hunting Bigfoot and Yeti since the 1950s, with little to show for it. In his usual upbeat and jaunty style, Byrne said he felt confident the creature would be found. Unfortunately, he continued, he feared the proof would probably not be brought in as a result of careful research and field work, but it would come in splattered all over the front grill of one of the huge logging trucks that sped with such seeming recklessness around the backwoods of the Pacific Northwest carrying felled trees to lumber mills to be turned into paper and cardboard boxes.[67] Byrne had invested too much in the hunt to give up so easily, however. His retirement was short-lived.

Greg Long recounts an interview he had with Patterson acquaintance Jerry Lee Merritt in 2003, who claimed Patterson had been loaned a 16mm camera from a local television station and used it to produce footage for his planned documentary. The footage he shot, according to Merritt, included scenes of some of his friends riding horses through the woods, leading pack animals, and similar horsy antics. One of the hammy cowboys was Bob Heironimus. [68]

In 1979 just one month after the announcement of his retirement, Peter Byrne wrote to advise Sydney Anderson that his examination of the Patterson film not only showed it to be a hoax, but that the version Patterson promoted was not quite as original as billed. Byrne said Patterson and company had made one version as a test, and then made an improved version for public consumption. The cowboys had "carefully viewed" the original for flaws and then "remade" a second version at Bluff Creek on October 20, 1967." He also warned Anderson about Patterson and Gimlin. They had, he said, run afoul of the law in the past.[69] This private correspondence never made it to the open. Not wanting anything to do with the monster hunters, Anderson just put the letters with the explosive accusations in his file and forgot about them. Byrne soon dropped such accusations. In discussions with Long, Byrne recounted the great lengths he had gone through to answer questions about how the film traveled from Bluff Creek to Yakima for the first screening in Al DeAtley's basement. He told Long that he finally managed to interview DeAtley, who told Byrne, "It's fake. I know it's fake."[70] Byrne never brought up his two-film theory to Long and does not seem to have mentioned it anywhere except in the correspondence with Anderson. A possible scenario would have Byrne hearing the same rumors as others had about Heironimus having a gorilla suit in his trunk as well as other whispers about Patterson having faked the film. One possibility is that Byrne may have been upset with someone over something, and in an emotional moment contacted Anderson. Later, after nerves calmed, Byrne realized the two film theory had no support and dropped it.

Byrne was not the only member of the faithful who had their thoughts about the Patterson film's possibly dubious nature. Bernard Heuvelmans never thought it genuine. While staying with Ivan Sanderson during the Minnesota Iceman incident, he had a chance to see the film. He thought Patterson and Gimlin believable enough, but felt the hair on the creature was all wrong, that it did not behave like hair on a living creature. He wrote in private notes that the whole thing was "an obnoxious hoax."[71] Although liking Grover Krantz and his work, and appreciating Krantz's years of effort to convince him otherwise, Heuvelmans remained adamant about the bogus nature of the film.[72]

In the end, Greg Long's book solved or did not solve the case, depending on which camp one occupied. In his cantankerous way, René Dahinden had hit upon the heart of the issue. Roger Patterson's character, whether ape or angel, could be separated from the film. It did not matter what others said about the film. Only the film itself mattered. Despite his lack of credentials, Dahinden had a natural ability to think like a scientist. In an interview with Greg Long, Dahinden summed up in his usual pithy way by saying, "Just examine the fucking film!...Fuck Al DeAtley! Fuck Roger Patterson! And fuck Bob Gimlin! OK? Ignore the human element. *Look at the fucking film!*" Then decide.[73]

## The Passing of a Giant

Ivan Sanderson helped lead the charge to legitimize the field of cryptozoology in general and anomalous primates in particular, and he acted as the first great promoter of the Patterson film. Like so many of the monster chasers, his marriage as well as his finances suffered because of his passion. His longtime wife, Alma Vioreta, initially supported his work and enjoyed the adventure of it. They loved each other dearly despite his often being away chasing monsters and investigating UFOs which, over time, took a physical as well as mental toll on them both. He once wrote her: "I missed you terribly, but I am utterly worn out and I am in real danger of losing my grip on continuity."[74] Some years later, when organizing his archives, Sanderson wrote on the bottom of this typed letter that he could not even remember what trip he had been on when he wrote it. Eventually, however, his career put too much of a strain on their lives and they parted.

A complex personality among a group of complex personalities, Sanderson always carried a certain disdain for the upper classes, whether intellectual or social. He did research into his family history and developed the peculiar hobby of tracking down other men who had the same name as him, collecting quite a few. Always troubled by his own parent's separation, he acquired their divorce papers, including the painful letters detailing the lawyer's discussions over whether his father wanted to pay his school tuition (his mother wanted to, his father said he could not afford to). At the same time, he saved yellowed and crumbling newspaper clippings of their wedding. He took pride in saying his family line descended not "from a bunch of Lowland nouveau-riches," but from "Picts and Heelanders; working stiffs; and we've been around unsoiled for a least 8000 years."[75] Still feeling self-conscious about his lack of credentials, Sanderson applied to his *alma mater* to remedy the situation. In November of 1968, as the Minnesota Iceman and Patterson film flap grew, Sanderson contacted the bursar of Trinity College, Cambridge, where he had done his undergraduate degree. He informed them that he had worked at Cambridge following his graduation in 1933 as a demonstrator in the zoology department and that he had done most of the work for the master's degree and had intended to go on for his doctorate in zoology when he headed off to do fieldwork and had never gone back to finish. He had, he told them, gone on to do substantial scientific work and publication in his field. With these circumstances in mind, he asked if Cambridge would award him the degree. To his great relief Cambridge said they would, and in 1969 Trinity College awarded him a master's degree.[76] He did not enjoy it for long.

In January of 1972, Alma passed away and at 61 years of age Ivan Sanderson took up with their mutual friend, Marion Fawcett, better known as Sabina. Acknowledging the unusual nature of their relationship, Sanderson told journalist Ralph Izzard (whom he had known since the early Tom Slick Yeti days) that Alma had wanted Sabina to step into her shoes and "take care of" her husband. Sabina and Ivan had much in common. He explained to Izzard simply, "we're both lonely and we love each other."[77] By November of that year, stomach cancer hospitalized him. After that he returned home one last time and died quietly on his farm in New Jersey on February 19, 1973.[78]

Disappointments dogged Ivan Sanderson's life. Each time he seemed close to proving the ABSMs real, just as he was about to grab one of the creatures by the scruff of the neck, something would come along to wrench it free to vanish before his very eyes. Even his papers were disrespected. His meticulous collections turned up burned or stolen, or just consumed by the vultures that descended on his farm after his death, carting away large masses of materials. He never received the respect for his efforts he felt was due him, certainly not from the bulk of the mainstream science community. Just a few years before he passed away, he told a relative in a statement of obviously exhausted defiance, "One day I'm going to have it out with the Almighty over this." Then he sighed, "I must say I would appreciate a break once in a while!"[79] The break never came.

Even in death, Sanderson could not escape ridicule. In 1976 William Montagna, director of the Oregon Regional Primate Research Center, published in his "From the Director's Desk" editorial column of the center's house organ, *Primate News*, a scathing critique of the Bigfoot phenomena in general and Sanderson in particular.[80] Montagna, who early on had been enthusiastic about anomalous primates, began by lamenting that schoolchildren often did not know the basics of science history, but could go on at length about Sasquatch. He pointed out that popular science publications like *National Geographic* regularly contained articles on the discovery of new species that drew little attention, yet interest in monsters had risen. He recounted how as director of the primate center, monster enthusiasts bombarded him with hair samples and footprint casts.

His annoyance growing, Montagna turned his attention to the Scot from New Jersey. He wrote that as Sanderson was working on *Abominable Snowmen* he approached Montagna to provide a supportive review. Montagna refused calling Sanderson a "distorter of facts," who had been "falsifying biological facts since his first useless, if fascinating book, *Animal Treasure*, was published in 1938." He called *Abominable Snowmen* "500 pages of gibberish...in execrable English." Montagna finished his column by saying, "I could not inform that colossal egomaniac that neither he nor his writing had improved with age."[81]

Ridiculed or ignored by the bulk of mainstream science, Ivan Sanderson still became an icon to cryptozoology enthusiasts around the world. He did enormous amounts of work uncovering legends and ancient writings describing such creatures. It is doubtful whether many of the scientists who saw the Patterson film would have done so if not urged on by Sanderson. His books and articles, especially *Abominable Snowmen* and the Bigfoot *Argosy* article, are prized by collectors who hold onto first editions with the passion of someone holding a First Folio of Shakespeare. He inspired most of the great names in the field, from Bernard Heuvelmans to Tom Slick, Roger Patterson, and Grover Krantz, and continues to do so for new generations of monster hunters.

## Conclusion

If Roger Patterson faked his film for financial gain, he failed. The money he made from the film while alive, which John Green referred to as "substantial," all went back into the hunt.[82] For all his personal flaws and dubious behavior, Patterson seems to have genuinely believed in Bigfoot. Patterson's story is just one of the endless tragedies of monster hunting. Whether he faked the film or whether it is genuine, it became far more trouble for him than it was worth. He became the first in a long line of monster hunters to succumb to cancer, dying in 1972. Robert Gimlin, quiet for so many years, finally began to give interviews and appear at Bigfoot conventions in the twenty-first century. Never really a committed enthusiast, he came to believe in, and insists today on the validity of, the film his friend Roger Patterson shot that day in October of 1967. He rejects the notion that Patterson hoaxed even him.[83]

Patterson's film has been studied by amateurs, evaluated by special effects experts, and looked at by scientists. It has been picked over, poked at, trumpeted as the most import piece of wildlife film ever taken, and laughed at as an obvious fake. The reason it resists scrutiny—and probably always will, regardless of whether it is genuine or fake—is the material fact that the film itself is of poor quality. Even with all the high tech gadgetry available to examine film, the low resolution of the original grainy 16mm footage renders it practically impossible to analyze in great detail. We may never know whether Patterson meant it to be this way, or that it was just the dumb luck of an individual unskilled and unsophisticated in the ways of filmmaking. In North America at least, it has become the toll booth all anomalous primate enthusiasts, academic or amateur, must pass to proceed. It lurks and skulks and peeps about just off to the side of every believer and skeptic, challenging, mocking, and encouraging. Regardless of who owns it, the Patterson film became a central component of Sasquatch studies. It allows for no middle ground. It is either real or fake, with no chance it is a misidentification of something else. Patty's now legendary backward glance in frame 352 teases and tests anyone who has ever seen it. It survives when all others associated with it have come and gone. It is not the only evidence, and it is not the only contentious evidence. Everything ever brought forward to support manlike monsters, mystery-apes, and anomalous primates has been controversial and will continue to be.

# Chapter 6

# The Problems of Evidence

*I don't expect much from such an orthodox anthropologist.*
Boris Porshnev on John Napier[1]

Contrary to the popular misunderstanding of the idea, a scientific theory is not a blind guess. It is a notion about how the universe operates, which is based upon a considerable body of factual and circumstantial evidence and ties those facts together. Without an explanatory theory a mountain of facts means little. In its simplest form, science works by gathering evidence and formulating theories. The problem Grover Krantz and other academic monster enthusiasts had stemmed from having theories, but not enough facts to support them, or at least facts the mainstream accepted. Superficially, the *Gigantopithecus* theory made sense, seemed logical, and could explain how such a creature came to be and how it came to inhabit the areas witnesses said it did. Monster hunters had three types of evidence in the form of eyewitness accounts, footprint casts, and photographs and films. Despite the physicality of the last two, all of these generated suspicion from a scientific point of view. In the absence of a Sasquatch body, Krantz tried to establish his theoretical work as best he could. He found fellow travelers in an unlikely place. The monster hunters of North America and England found allies in Russia, where similar creatures, commonly called Almasti, had been reported for years, drawing the attention of a group composed of both academic and amateur investigators. The Russians, however, employed a very different explanatory theory.

## The Importance of Being Erectus

Like any good scientist, Grover Krantz worked to build up layers of theoretical foundations to support his overall thesis that

*Gigantopithecus* represented the progenitor of Bigfoot. Applying a form of multiregional or continuity theory to the explanation for Sasquatch allowed Krantz to have a sound theory to explain the creature's existence. His views on the history of *Homo erectus* are crucial to understanding how Krantz came to accept continuity. If *Homo erectus* had the history Krantz believed it did, his Sasquatch theory made that much more sense. In Krantz's view the evolutionary history of Sasquatch began with *Homo erectus*, which branched off from a common ancestor—probably *Australopithecus*. Once they had evolved into their new form, *Homo erectus* populations then spread around and out of Africa. As they roamed about the globe the different environmental pressures they encountered forced them to continue evolving. It is in this transition stage, Krantz believed, that while they had yet to achieve modern human status, these *erectus* populations began to take on some of the characteristics of later human races. Finding homes in various parts of the world, *erectus* populations continued to evolve until they had become *Homo sapiens*. In this way they were all subspecies of *Homo sapiens*, all related, and all human regardless of being African, Asian, or European. Krantz went one step further; he added at least one additional lineage to *Homo erectus*.[2]

While all this branching occurred during the *erectus* phase, another population, responding to yet another set of environmental pressures in Asia, evolved from *Homo erectus* into the larger form of *Gigantopithecus*. This population had not achieved the humanlike qualities of the other erectus groups. They remained closer to their primate antecedents. They retained bipedal stance, but did not acquire rudimentary speech or higher cognitive abilities like their cousins. They veered away from the path toward human and increased in their overall bulk. As is the nonstop pace of evolution, at least one *Gigantopithecus* population, likely in the region of the Himalayan Mountains, evolved into the Yeti. Then either another *Gigantopithecus* population spread into the Americas and became the Sasquatch or a Yeti population moved further east to become Sasquatch. This was the chain. This history made Sasquatch a relation to *Homo sapiens* but not an ancestor. If Krantz could articulate this point persuasively he could convince his colleagues. He delved deeply into *erectus* studies and from his student days at Berkeley began producing scholarship on the subject. He laid a foundation of *erectus* publications and data which he later could draw upon to explain Bigfoot.[3] It did not prove quite so easy.

Mainstream anthropology, the primary audience for Krantz's work, either dismissed it or thought his methodology questionable.

Indeed, the *Gigantopithecus* theory did not even become the standard within the monster community. A scholarly reviewer called his *Climatic Races and Descent Groups* (in which he argued that modern human racial traits first appeared in *Homo erectus* populations) "an idiosyncratic skewing of the questions being faced in physical anthropology." As for Krantz's idea about language propelling *Homo erectus* evolution, the reviewer wrote, "The use of data and the analysis go beyond the bounds acceptable to most."[4] Another reviewer thought it unusual to consider fingerprint patterns, cephalic index, and dental anomalies on a par with genetic traits, as Krantz did. He wanted to take the lead in Sasquatch research, but Krantz had fallen woefully behind the latest work on genetics and their application to questions of hominid evolution and diversity. Reading Krantz's book, the reviewer said, made him feel as if "three decades of thought and research [on genetics] had never transpired" and that the book showed "an absence of knowledge about or an outright rejection of truly vast realms of theoretical and applied human population genetics."[5]

Despite their friendship, Milford Wolpoff, along with Alan Thorne and their colleagues, responded negatively to Krantz as well. They claimed Krantz, like many others, did not really understand how regional continuity worked, adding, "We think it is a mistake to propose that there must be some intermediate position" between continuity and replacement theory, as Krantz proposed.[6] Another problem for Krantz's published works is that they show a lack of rigorous footnoting and only brief bibliographies often devoid of the latest sources. He suffered from this same problem since his student days. One of Krantz's anthropology professors at Berkeley warned him of his writing issues, which included making assertions without supporting them with cited evidence. The professor noted: "You would certainly strengthen your presentation of ideas if you would illustrate them with apt examples." Krantz referred to himself in his papers with expressions such as "I think" this or that. In one paper he said, "Just because others don't think clearly doesn't give me an excuse not to." In a number of his papers he dwells on the issue of definitions of well known ideas, to which the professor replied "Why not consult a good dictionary?" if he did not understand the meaning.[7]

Using *Gigantopithecus* as a model could also be problematic. Krantz said the dentition of *Gigantopithecus* was similar to *Australopithecus afarensis* from Africa, therefore if *Australopithecus* was a biped, the same could be said of *Gigantopithecus*. "This creature [*Gigantopithecus*] is no more apelike than *afarensis*," he said, "which is indisputably bipedal."[8] In the first description of the

Indian *Gigantopithecus bilaspurensis*, E. L. Simons and S. R. K. Chopra said: "In various ways the new specimen resembles species of *Australopithecus....*"[9] Krantz reasoned that if *Gigantopithecus* was a biped, so was Bigfoot. Complicating things for Krantz, members of the paleoanthropology community were divided over the nature of *Gigantopithecus.* In 1956 Pei Wenzhong (1904–82), considered with Jia Lanpo (1908–2001) as one of the fathers of paleoanthropology in China, said of *Gigantopithecus*, "The basic nature of this animal was that of an ape."[10] Yet Pei also argued, along with Franz Weidenreich, that the name should have been *Gigantanthropus* and believed the creature was over twelve feet tall and "approaching the status of man."[11] Yale paleoanthropologist, David Pilbeam, and a number of others believed the dentition of *Gigantopithecus* showed it more accurately to be a pongid: an ancestor to the orangutan. Pilbeam argued that just because *Gigantopithecus* had humanlike teeth did not mean it was necessarily related to the human line.[12] In 1986 California State University anthropologist Bruce Gelvin stated, "*Gigantopithecus* is an extinct side branch of the *Hominidae.*"[13] Just how this discussion over the place of *Gigantopithecus* in the hominid line would have changed Krantz's position is unclear. He does not seem to have bothered to take what other paleoanthropologists said into account about the nature of *Gigantopithecus.*

## Supporting the Theory

While he never articulated it in these terms, Krantz posed a series of questions that had to be answered in order to solve the Bigfoot mystery. First, did fossil evidence of a Sasquatch-like progenitor exist? Second, how could these creatures have made their way into North America? Third, what would account for the diversity of anomalous primates of varying sizes and colors reported around the world? Fourth, could the laws of biomechanics account for such a creature? Finally, did any evidence prove the animal's existence today? He believed he had the answers. The progenitor was *Gigantopithecus*, the Bering land bridge allowed it to enter the Americas, regional continuity—or at least something like it—accounted for the diversity, and the Patterson film and footprint casts proved the creature still alive and active.

Krantz followed the physical evidence he had before him. While he, too, sensed shenanigans had been going on at Bossburg and Colville, the footprints kept nagging at him. John Napier, too, was taken with the Cripplefoot. He saw a specific pathology to the print: clubfoot or *talipes-equino-varus.* This condition created just the sort of malformation he saw in the track. Both Napier and Krantz concluded this deformity was the result of an injury. They also both agreed that the track could not have been faked. "It is very difficult," Napier argued, "to conceive of a hoaxer so subtle, so knowledgeable—and so sick—who would deliberately fake a footprint of this nature."[14]

Convinced the Bossburg tracks were genuine, Krantz received word of a new set of particularly convincing tracks. In June of 1982 U.S. Forest Service worker Paul Freeman (1943–2003) found tracks at Mill Creek in the Blue Mountains near the city of Walla Walla, Washington. His job as a patrol rider included checking the area—the watershed for Walla Walla—on a regular basis. This is how he came across the tracks. While believed by some, including Grover Krantz, many saw Freeman as a fraud. The ridicule became so great, he said, it forced him to leave his government job to study the phenomena more closely. Freeman seemed a carbon copy of Ivan Marx. He, too, had an uncanny ability to find Sasquatch evidence. In the late 1980s one of Krantz's WSU graduate students, Lonnie Somer, acquired alleged Bigfoot hair to examine. The hair samples came from a location Somer found through Paul Freeman. Somer published his results in an article for the journal *Cryptozoology* in 1987 and presented them publicly at a conference held at WSU in 1989. Somer's conclusion: the hair provided by Freeman was synthetic and consistent with a wig.[15] Once again, unusual evidence had been eliminated by standard scientific inquiry. Still, Grover Krantz would not be deterred.

## A Monster by Any Other Name

Sasquatch and its kin had many popular names around the world, but no scientific designation. Krantz wanted to remedy that. At the January 1985 meeting of the International Society of Cryptozoology held at Sussex, England, he presented the paper, "A Species Named from Footprints." He felt this "should serve to structure further inquiry into this matter along sober lines, and to discredit some of the unfounded speculation."[16] Acutely aware of the perception that many in the mainstream considered anomalous primate studies pseudoscience, Krantz wanted to distance his work from what he considered the lunatic fringe by keeping it within the bounds of accepted methodological science. The usual procedure in the description of a new species included assigning a binomial designation to it.

Krantz began his Sussex conference presentation by explaining the nature of *Gigantopithecus*. He argued that as *Gigantopithecus* existed and conformed to the Sasquatch evidence it was reasonable to assume one supported the other with an evolutionary connection. While not mentioning Bernard Heuvelmans's *Gigantopithecus* theory for the Yeti, he did give credit to John Green for applying it to Sasquatch. Krantz admitted he stood on shaky ground with this methodology, but felt it a way to legitimize something his anthropology colleagues thought at best a piece of folklore and at worst an outright hoax. Some eyebrows did rise at the realization the evidence he used to support his bid to name the creature took the form of casts made from the footprints found at Mill Creek by Paul Freeman. Despite the origins and nature of his evidence, Krantz said, "I fail to see a good reason why this should make any difference in their acceptability" as type specimens.[17] Krantz's first choice for the animal was to simply call it *Gigantopithecus blacki*. If a relative of this creature, he proposed the name *Gigantopithecus canadensis*; if an *Australopithecine*, it should be called *Australopithecus canadensis*. (In 1974 Krantz's Russian anthropologist friend, Boris Porshnev, proposed the name *Troglodytes recens*.)[18] Bernard Heuvelmans told Krantz he should have named it *Sasquatchicus krantzii*.[19]

Krantz's bid to name a monster was not without precedent. There had been a somewhat successful monster naming attempt ten years before. British naturalist Sir Peter Scott and American lawyer/engineer Robert Rines put forward the name *Nessiteras rhombopteryx* for the famous monster of Scotland's Loch Ness. Rines had taken a group of underwater photos of the creature, which had attracted much attention, even rating an article in *Nature*.[20] Ten years after Krantz's attempt, marine biologist Edward Bousefield argued for the scientific name of *Cadborosaurus willsi* for the lake monster from British Columbia.[21] These attempts to name anomalous creatures found their basis in the argument that they were related to known fossil forms (*plesiosaur* for Loch Ness and the fossil whale *zueglodon* for *Cadborosaurus*). As Krantz, too, used an established fossil species as a basis, he assumed he too would be successful.

Following the conference, Krantz sent his paper to editor Leigh Van Valen (1935–2010) at the University of Chicago Press, hoping to get it published.[22] Van Valen, a friend and ally of Krantz, sent the paper out to a pair of scholars for blind reviews, but only one response survives. This reviewer turned the paper down for publication, saying that naming a species "conditionally," as Krantz suggested, caused complications, and found fault with the notion that "it is based entirely on the assumption that some crucial traits will be similar to living apes and that a mandible with horizontal rami means bipedality." Krantz argued that the shape of the

*Gigantopithecus* jaw indicated bipedalism. The reviewer pointed out that many living apes have similar jaw morphology, but are not bipeds. The reviewer also could not resist looking for other reasons for the attempt to name the creature, suggesting that Krantz wanted to name Bigfoot before the creature was proven to exist "in the hope that if it should turn out to be correct, he will be recognized as a prophet."[23] In their extensive late–twentieth-century work on *Gigantopithecus* Russell Ciochon, John Olsen, and Jamie James said Krantz had "stepped outside the bounds of science: zoological names are always assigned to type specimens...something which is lacking in the case of Sasquatch."[24] Krantz argued foolproof evidence did exist to make these casts type specimens from a living animal, and that he had been the first to see it.

## Dermal Ridges

Krantz said that any hoaxer "had to outclass me...and I don't think anyone outclasses me...at least not since Leonardo da Vinci."[25] In *Bigfoot/Sasquatch Evidence* (1999), Krantz argued that a hoaxer would have to possess a special prowess, "more so than me, and I seriously doubt that any such person exists."[26] Krantz wanted to establish a narrative to support his position as academic leader of the field, separating him from the amateurs. This narrative held that only a trained academic had the anatomical knowledge needed to distinguish genuine from fake prints. Therefore, the amateurs did not have the expertise or training for a proper study of the creature. To emphasis this point, he discovered an anatomical detail that could be neither faked by, nor recognized by, an amateur. What Krantz had seen, first in the Cripplefoot tracks, then in the ones used as type specimens as well as others, came in the form of tiny, barely perceptible ridges in the footprint casts. Closer examination of these structures seemed to show them to be dermal ridges, the foot's equivalent to fingerprints. Krantz saw this as major discovery, one which could not possibly be ignored by mainstream anthropology.[27]

Unfortunately, Krantz was wrong about the mainstream. In the 1980s and 1990s skeptic Michael Dennet argued that most of the footprint casts Krantz and others counted as genuine had been faked. This meant the dermal ridges must be fake as well, or had appeared unintentionally, but were mistakenly interpreted by Krantz as dermal ridges. Dennet interviewed other scientists who thought Krantz mistaken in his assessment of the "detailed microscopic anatomy" he claimed to have found in the dermal ridges. [28] He also pointed out that Krantz's two primary suppliers of casts, Ivan Marx and Paul Freeman, had hoaxed, or stood accused of hoaxing, prints. Krantz's adversary, René Dahinden, lost faith in him when they clashed over the Cripplefoot tracks having dermal ridges. Dahinden questioned anything associated with Marx. He was upset Krantz had made them the centerpiece of his argument for the existence of Bigfoot. If he continued to do so, Dahinden told Krantz, "I will be on your ass every inch of the way!"[29]

Continuing his program to make his Bigfoot research empirically based, Krantz focused on the footprints. He discovered the dermal ridges while making a latex copy of the track casts given him by Ivan Marx. An accident of circumstance had Krantz using two different colors of latex—white and red—to make the copy. The contrast of the two colors allowed him to see this effect for the first time. He excitedly sent out his findings to a number of anthropologists for review. He also hit upon the novel idea of sending the dermal ridge evidence to law enforcement personnel, specialists in fingerprinting techniques. While a few of the anthropologists showed mild interest, most discounted the very notion out of hand. He had better luck with the nonacademic police and sheriff officers. Several fingerprint experts deemed the dermal ridges real and proof of the existence of the creature in question. One fingerprint expert in particular, law enforcement veteran Ed Palma, considered the dermal ridges significant. He tried to get the Nike sneaker company to examine the casts. A Nike spokesperson gently declined the offer. He did tell Palma, however, that Nike had made custom sports shoes for athletes with sizes similar to the Bigfoot casts.[30] Ripu Singh of the University of Windsor, Ontario, thought the dermal ridges intriguing. With his graduate students assisting, Singh studied the casts closely. He was puzzled, though, by the fact that he did not see the usual dermal ridge patterns like cores and deltas that should be there.[31] Professor A. G. de Wilde of the University of Groningen, The Netherlands, looked at the material Krantz sent him and declared the prints "are not from some dead object with ridges in it, but come from a living object able to spread its toes." He also said the prints came either from a human or a humanlike creature.[32] Krantz tried to get everyone and anyone, from the FBI to Scotland Yard, to look at the dermal prints. Few showed anything more than mild interest. John Berry, a fingerprint expert and editor of the fingerprint community's journal, *Fingerprint Whorld*, told Krantz Scotland Yard thought the prints probably real.[33] Krantz even had a palm reader look at the prints.[34]

Despite all his efforts, the dermal ridges just did not get the kind of professional attention Krantz felt they deserved. He wrote an extensive article on the ridges for the journal *Cryptozoology*, but it drew no attention. Richard Greenwell, secretary of the society, told Alex Roche, president of the American Dermatoglyphics Association, that he could not get anyone outside the cryptozoology community to examine Krantz's article.[35] Later, John Berry along with Stephen Haylock, published "The Sasquatch Foot Casts," in *Fingerprint Whorld*.[36] Their article drew little more attention than Krantz's had.

The history of anomalous primate studies spills over with examples of evidence that looked convincing. Everything from the Patterson film to the Cripplefoot tracks, to the Minnesota Iceman and dermal ridges, has been held up initially as beyond the ability of hoaxers to create. Like Piltdown Man before them, these artifacts have been viewed in different lights by different viewers, all of whom saw what they were looking for. What is often forgotten in the "hoaxers could not do this" argument is that too much emphasis is placed on what hoaxers could or could not do. In the end, hoaxers do not really need to do much at all. That way, someone looking at the artifact will fill in the blanks themselves. No super cunning or technical expertise is needed by a faker of evidence. All they need to do is throw something together and let their audience do the rest. Once some detail is seen, later viewers will likely see the same. Then, when skeptics point out the flaws, they will be labeled by believers as blind, closed minded, unwilling to think outside the box, or stooges of the academic power structure or an evolutionary mafia. The original hoaxer then simply steps back and watches the crackpots and the eggheads hash it out among themselves. No further work is required except to chuckle at the havoc they have wrought.

## The Morphology of a Bigfoot Encounter

While little consistency existed in physical monster evidence, a number of meta-ideas about the field in general could be articulated. The different types of evidence for manlike monsters have suggested to enthusiasts that a number of logical narratives could be employed as unassailable paradigms. Some of the most popular are:

- Sasquatch and other manlike monsters from around the world are genuine biological realities, not figments of the imagination
- The vast landscape of North America or Asia, or any of the wild places of the world, make it easy for these creatures to hide themselves

- No physical remains have been found because scavengers and other environmental effects eliminate any traces of a dead animal
- So many eyewitnesses cannot all be wrong
- Scientists do not believe these creature real because they are too wedded to scientific theory and have no real world experience of the creatures

Not immune to narrative building, skeptics had their standard explanations as well.

- Bigfoot exists only as a figment of the imagination
- Bigfoot sightings are either hoaxes or misidentifications of other animals
- Bigfoot comes under the heading of the paranormal and so cannot be counted a legitimate field of scientific investigation
- Monster enthusiasts with no scientific training have no idea what they are doing

Such divergent views made it difficult for the two sides to come together. The many examples of different evidence had itself become an issue. Not all the footprint casts had dermal ridges, not all the feet had uniform structure. Some seemed to show characteristics that others did not. Most had the standard mammalian five toes, while some showed only three. There never seemed to be footprints from juveniles or infants, only grown adults. One realm that did show a sense of uniformity came in the eyewitness accounts.

Embedded in the largest body of evidence for manlike monsters, a certain pattern can be discerned. Supporters held that, taken in the aggregate, the sheer bulk of eyewitness reports supported the reality of ABSMs. Hundreds of sighting occurred every year. The caliber of some eyewitnesses—park rangers, police officers, and people with nothing to gain from fabricating stories—seemed to give an air of legitimacy to them. The historical nature of accounts going back to the nineteenth century, not only in Native American lore but in written accounts of Euro-American and Canadian traveler's tales had a reality of its own. While supporters, including Grover Krantz, admitted that not all stories could be taken at face value, and that many likely fell into the category of misidentification, enough remained to constitute an solid body of work. A standard practice of investigators had them compiling file cabinets full of reports they gathered directly from the eyewitnesses themselves, often shortly after the encounter. Investigators also scoured diaries, newspaper articles, and published works from early explorers, looking for any reference to such creatures. Indeed, the bulk of most books on anomalous primates, and cryptozoology in general—beginning with Ivan Sanderson and Bernard Heuvelmans—are compendiums of eyewitness reports.[37]

Eyewitness reports of manlike monsters can be viewed as a body of literature and are often read as scripts. A generic mystery-ape encounter story goes like this: a lone eyewitness is traveling through a rural environment. They round a bend in the trail or road they are on and suddenly encounter a strange creature up ahead of them. The creature is usually at a distance of some yards and has just stepped out of the brush onto the road the eyewitness is on, or it has just stepped off the road. The creature is an upright biped. It is hairy and, while apelike, gives off the distinct aura of a man, especially in its facial features. The creature stops momentarily and looks at the eyewitness, then disappears into the brush never to be seen again. The eyewitness does not have a camera or a firearm—when eyewitnesses like hunters who are armed have this experience they often hesitate to fire because of the creature's vaguely human appearance. The eyewitness is left shaken. All this occurs in less than 30 seconds.

During these encounters the creature, with only a handful of exceptions, does not approach the eyewitness, and the eyewitness rarely if ever pursues the creature. The creature never seems startled by encountering a human; in fact, descriptions of the animal's behavior during the encounter are almost blasé, as if it sees humans on a regular basis. Even when the eyewitness drives a vehicle at night the animal does not show the usual characteristic of the deer caught-in-the-headlights most wild animals exhibit in such situations. The creatures, like lightning, never seem to strike twice. Once observed crossing a road, for example, they never seem to take that trail again. No Sasquatch or Yeti trails or annual migration routes have been identified. There are no predictable movements or behavior, no known watering holes or areas where the creatures regularly eat or visit.

Eyewitnesses rarely have cameras with them or when they do get only blurry indistinct shots. The creatures are rarely seen by large groups of people at a time. They normally are one-on-one encounters with only a few sightings of more than one creature at a time and rarely, if ever, of juveniles or infants. With the exception of the Bossburg incident, rural villages and towns rarely see the creatures enter their precincts—like the polar bears of Churchill, Alaska, do so regularly that tourists travel there in numbers to see and photograph them. The facial features of the creatures are universally said to be humanlike, even showing intelligence and human understanding. The eyes are expressive and are never described as cold or lifeless the way a shark's eyes are always described.

All this seeing of anomalous primates has yet to produce physical evidence other than footprints. What sort of evidence would be of use? Models of unknown hominid evidence can be taken from what comes from the fossil record. *Gigantopithecus*, the heart of Grover Krantz's argument, is known from little evidence: only teeth and jaw parts and nothing else. Yet it is accepted as a fossil primate. Hominids like Neanderthals and *Homo erectus* have substantial fossil records, while *Homo habilus* and the *Australopithecines* are relatively scanty.[38] To prove Sasquatch and Yeti real like these species, a tooth or two or a genuine bone fragment would secure the entire thing. The find would have to be carefully documented the way any fossil find or endangered species would. Turning up in a refrigerator in someone's garage would likely fall short of the mark, unless it included substantial biological material. Any evidence would need the imprimatur of a respected museum or university.

The morphology of a Sasquatch encounter described above comes from sightings in North America. However, similar encounter stories and evidence have come from the other place in the world with a high level of ABSM activity and a body of researchers equally passionate about making the creatures real. While agreeing on a number of points about the nature of anomalous primates, these researchers have taken a very different approach to explaining their origins.

## The Russians

In 1958 the Soviet Academy of Science, convinced the reports of the Yeti and other Central Asian Snowmen had at least some merit, sponsored an expedition to the Pamir Mountains to look for them. These sightings generated such interest; a newspaper reported "a special Soviet collective has been organized here to track down the Snowman once and for all."[39] Since Western-led expeditions to the Himalayas searched for the Yeti, the Soviets believed they should as well. Action by the state under a newly created Snowman Commission allowed the operation to be quickly assembled—staffed by genuine scholars—and well equipped.[40] Reports of sightings by geologist A. G. Pronin, who spotted a Snowman while on a state sponsored geological expedition, had contributed to the official attention. "I had heard reports of these creatures," he said, "but I never expected to see one myself."[41] In Russia manlike monsters are generally referred to as Almasti, while in Mongolia they are called Almas. Conforming to the basic outline of anomalous primate, the Almasti/Almas have cultural and morphological differences distinguishing them from other members of the group. Also, the Russian monster hunters tend to object to the use of the term anomalous "primate" as well as "monster," as they tend to see the creatures as more human than ape. They prefer to call these creatures *Homins*.

As with so many monster expeditions, the Pamir trip found nothing despite the government's sanction. As a result of the expedition's poor performance, the Soviet state withdrew its funding, ending the brief life of the Snowman Commission. The Soviet press spun the story, arguing that Communist science had shown Snowmen did not exist and, thus, Western expeditions to the region were just thinly veiled spying activity and sinister capitalist intrigues. A state publication on the expedition brought outside attention not only to the existence of the expedition itself, but also to the Almasti question.[42] As in the West, in the East monsters found themselves pursued by a small cadre of academics and amateurs rather than organized mainstream science. Like their Western counterparts, these researchers argued that the creatures did exist, and they set out to prove it.

### Boris Porshnev

The man most responsible for the Russian Snowman expedition to the Pamir Mountains, Boris Porshnev, had widely ranging interests. Alternately described as an historian (of French revolutionary history), a social psychologist, and an economist, Porshnev (1905–72) found himself fascinated by anomalous primates as well.[43] Not a trained anthropologist, Porshnev approached the origin of humans from a social psychology studies perspective. Less interested in the physical remains of human ancestors, he sought the intangible qualities of humanness: creativity, family organization, behavior, and brain function. He thought to place human development into a Marxist perspective, yet with a humanist twist. As did some Western scientists, Porshnev thought answering questions about the existence of the Almasti and Sasquatch would illuminate important aspects of human social growth.[44]

The founders of modern socialist thought, Karl Marx (1818–83) and Frederick Engels (1820–95), both incorporated a reading of Darwinian evolutionary thinking into their explanations of class struggle (this in turn spawned a debate over the supposed role of Darwin in the founding of Communism that rages to this day). They argued that humans were neither divinely created nor simple brainless automatons reacting instinctively to outward stimuli. Engels supported the notion that in the distant past a group of anthropoids began the trudge towards humanness by standing up and using their hands to manipulate tools, thus creating labor, which in turn led to the natural condition of a socialist society. These early human ancestors lived in harmonious union with each other because of the invention of work. Building upon the writings of American anthropologist Lewis Henry Morgan (1818–81) on the origins of the family unit, Engels argued that "the monogamous patriarchal family—like Capitalism—is not a fixed feature of human society in all ages, but the product of historical development."[45] In other words, socialism can be traced back to man's earliest days, prompted by the changeover to bipedal locomotion, the invention of work, labor, and communal living, and represents man's natural state. Capitalism, on the other hand, arrived as a later and unnatural development. Supporting this idea, Russian biologist, and primate hand specialist, Professor L. Astanin told Grover Krantz that Bigfoot's hand had the structure it did because "in its evolution [Bigfoot] did not take to labor and thus retained the primitive proportions of the hand."[46] It therefore did not rate a place in direct human lineage the way Almasti did.

Boris Porshnev saw monster studies as a way to confirm the evolutionary and materialistic explanation of human behavior. In this way he sought to start a revolution in paleoanthropology the way Marx, Engels, and Lenin had started one in politics. But resistance to unusual data transcends political and cultural borders. Porshnev did not belong to the Communist Party. Indeed, he found himself on the receiving end of Party ridicule, not only for his Snowman work, but for his work on the French Revolution because of his less strident humanist take. His books *Social Psychology and History* (1970) and *Leninist Theory of Revolution and Social Psychology* (n.d.) focused on the earlier Marxist-Leninist approach to the subject rather than the more brutal Stalinist way. As so many academics in Soviet Russia had to do at the time, Porshnev walked a precarious line between science and the state.

In addition to the ever present threat of state sanction, Porshnev found himself with a bit of a conundrum. The Almasti seemed distinctly different from the Yeti and Sasquatch, which did not offer any help for explaining social organization. He read Heuvelmans and Sanderson and agreed the Yeti likely descended from the apelike *Gigantopithecus*. The Almasti posed a different scenario altogether. Like other anomalous primates, the Almasti and Almas had a mythological history. As the wild men of Asia, stories of their existence went back centuries in local folklore. Unlike their bigger cousins, the Almasti had a history of benign social interaction with the peasantry.

They did not have violent reputations, and did not present a monstrous aspect. Indeed, the Almas usually had to be wary of humans rather than the other way around. They differed from the Yeti in a number of other respects. The Almasti had no apparent mystical or religious baggage. They did not have the physical stature of the larger Yeti, being more manlike in their proportions. Their hair was shorter, but matted and mangy. They did not seem as majestic and awe-inspiring; in fact, they seemed mongrels by comparison, giving off an almost groveling, beaten appearance. All this suggested to Boris Porshnev and other Russian monster enthusiasts that the Almasti constituted the remnants of a population of Neanderthals that had survived into modern times as relics of an important but bygone age.[47] This direct connection to the human past made the Almasti a good way to examine the human condition in its earliest stages and thus support the Marxist interpretation of human evolution. If they could be proven to have existed, and better yet to still exist, it would be a great confirmation of the socialist approach.

Porshnev first learned of Almas through the work of ethnologist Badzar Baradiin (d.1940), who in the early twentieth century had been collecting cultural artifacts and legends of his Mongolian homeland. Porshnev learned of Baradiin's work through another Mongolian scholar, Yöngsiyebü Rinčen (1905–77).[48] According to the popular account, Baradiin (sometimes spelled Baradyine) saw an Alma while exploring Tibet in 1906. Baradiin was then accosted by a state official, who convinced him not to publish his findings. Instead, Baradiin told his story to a colleague, Tsyben Žamcarano, who quietly began research into the creatures on his own, including gathering eyewitness reports, dates of sightings, and map locations. Žamcarano then shared his work with academician Rinčen. Unfortunately, Žamcarano's archive vanished into the Byzantine labyrinth of Soviet officialdom. Porshnev tried to find this archive but it had either disappeared or the state denied his requests to see it.[49]

This is the general story as related by most authors on the subject, including British archaeologist Myra Shackley, who wrote extensively on Russian monster hunting.[50] In 1983 Michael Heaney, a Russia scholar at the University of Oxford's Bodleian Library, revisited the story.[51] At the time of his explorations Žamcarano held a position of relative power in the Soviet satellite Mongolian People's Republic, having been instrumental in setting up the country's science academy. As was the case in those days, one could be in state favor one moment and out the next. Žamcarano had a deep affection for his native Mongolian culture and collected and promoted Mongolian history. The Soviet government, however, did not want its colonized people showing any self determination, instead promoting the idea of a Soviet hegemony. As a result Žamcarano found himself labeled a "bourgeois nationalist," Soviet code for an ethnic trouble maker. In 1937 he was thrown into prison, where he died, probably in 1940.[52] The colleague who started him on his quest to find Almas, Badzar Baradiin, suffered the same fate for his pro-Mongolian activities. Like the vanishing commissars—those high ranking officials who had fallen out of favor with Stalin and been literally erased from official photographs—Baradiin and Žamcarano became vanishing scientists.

By the 1960s academician Rinčen found himself in a tight situation as well, when he published a finding aid to Žamcarano's papers.[53] Yöngsiyebü Rinčen was one of the more unusual characters in Russian monster hunting. A Mongolian linguist and ethnologist—who like Baradiin and Žamcarano had an affection for his homeland and culture—he, too, had an interest in the Almas. He studied several skulls found in the mountains purported to be Almas.[54] Along with others in the monster community, he published in Carrado Gini's journal Genus.[55] Myra Shackley referred to him as "the eccentric and enigmatic Professor Rinchen whose chequered career and colorful personality made him persona non grata in Mongolia during some periods of his life."[56] By the early 1960s the political winds changed in Russia and Žamcarano's career came back into favor. It was during this period that Rinčen gained access to his papers and published his finding aide. In the convoluted bureaucratic thinking of the cold war, Soviet authorities reshuffled the papers so Rinčen's finding aide would be useless to outside researchers, effectively burying Žamcarano's work.[57]

In 1982, Oxford scholar Michael Heaney gained partial access to Žamcarano's papers in Leningrad (now St. Petersburg), Russia. To his surprise Heaney found no references to Almas "nor is there any evidence in his diaries or collected materials that he collected tales about Almas."[58] There seems to be no mention of an Almas sighting in the papers of Baradiin either. A number of details Porshnev claimed to have taken from Baradiin's writings also are absent. The details of Almas sightings Porshnev refers to are from the novel The Ravine of the Almases by Soviet journalist Mikhail Rosenfel'd. When he found Žamcarano's paper's academician Rinčen does not seem to have told Porshnev their location. A pro-Mongolian not especially enamored of the Soviet state, Rinčen may have intentionally steered Porshnev towards the Rosenfel'd novel and away from the "lost" archive of Žamcarano. Heaney's research showed the Baradiin sighting and the

Žamcarano research do not exist or, if they do, are so hidden as to be effectively nonexistent so they could not have been utilized by Porshnev.

## The Porshnev Theory

Boris Porshnev may have relied on faulty knowledge of the Baradiin sighting, but he used other evidence as well. The sighting he put most stock in came from Leningrad University hydrologist Alexander Pronin, who claimed to have seen an Almasti while on an expedition to the Pamir Mountains in the summer of 1957.[59] The Pamirs are part of the Himalayas and Hindu Kush and stretch from Tajikistan and Kyrgyzstan through China, Pakistan, and Afghanistan. They contain the Fedchenko Glacier—where Pronin saw his monster—which is the longest such feature in the world outside the polar regions. It is a beautiful, but cold and forbidding place.

Porshnev championed the notion of surviving, or relic, populations based upon his reading of the literature then current in the paleoanthropological record. Discussions of *Homo* fossils just then being discovered in Africa, as well as knowledge of Neanderthal anatomy, led him to accept that small groups of Neanderthals had survived into historic times. The prevailing wisdom of mid–twentieth-century paleoanthropology held that human ancestors appeared and went extinct in a simple linear fashion, with proceeding species disappearing as new ones appeared. Porshnev rejected this notion and argued that there may have been cases where previous archaic groups held on in small pockets after the bulk of their population had gone extinct. By the latter part of the century the simple linear pattern had been dropped by the mainstream. Continuing fossil finds gave a greater and more nuanced view of human evolution as a highly complex affair. The mainstream view now had human evolution as something where wide ranging hominid diversity had numerous *Homo* species coexisting simultaneously. Some paleoanthropologists argued that there may have been times when as many as a dozen *Homo* species coexisted.[60] Porshnev differed from this position by arguing that some groups, most notably the Neanderthals, did not just coexist with *Homo sapiens*, but continued to exist. The mainstream rejected this idea, arguing the fossil evidence showed that by about fifty thousand years ago only one hominid group, *Homo sapiens*, still lived.[61]

Porshnev took heart and influence from the work of Ivan Sanderson and Bernard Heuvelmans, who used the Neanderthal relic idea to explain some ABSMs. As a historian of French history, Porshnev likely encountered Heuvelmans's *Sur la Piste des Bêtes Ignorées* in its original French edition. Porshnev never accepted direct primate ancestry for humans. It did not quite fit his Marxist position. He used something more in line with the Pre-Sapiens theory, from the early twentieth century, of a humanlike ancestor going deep into the past with little or no direct connection to apes. Rather than separating them out into different groups, Porshnev placed all nonprimate, nonmodern human ancestors into the *Troglodytidae*.[62] He took the term from classical writers, who had commented on the existence of what Porshnev took to be Almasti-like creatures. The Almasti, he said, represented a relic of man's troglodytic past. Porshnev read ancient authors such as Pliny and their references to satyrs and wild men as descriptions of relic Neanderthals. When discussing why the Pamir expedition looked for Almasti, he said Almasti "could be classed as Neanderthal, but only anatomically and not in regard to their way of life."[63] Despite his affection for the Almasti, Porshnev saw them as a primitive, almost degenerate, form of protohuman rather than genuine human. They stood just close enough to humans to show their early socialist beginnings, which made them a valid topic of research. The relic theory also implied some of these relics might still be alive.

### Zana

His belief in the relic Neanderthal theory to explain Almasti made Boris Porshnev a natural ally to Bernard Heuvelmans. When shown evidence of the Minnesota Iceman, Porshnev immediately agreed with Heuvelmans that this creature was not a hoax and certainly not a Yeti or Sasquatch, but a Neanderthal. The Iceman convinced Porshnev even more that his relic theory had merit. The two men felt so strongly about the relic theory, and the Iceman's role in it, they collaborated on what is still the most thorough discussion of ABSM anatomy ever produced, *L'Homme De Néanderthal est Toujours Vivant* (1974).[64] Far more detailed than Ivan Sanderson's magazine article on the Iceman, Heuvelmans and Porshnev used the creature as a vehicle by which to promote and prove the relic theory.

Along with their discussion of the Minnesota Iceman, Heuvelmans and Porshnev also discussed the curious case of the legend of "Zana."[65] Local stories and folklore held that a live Almasti had been captured and held in human society for years. A colleague of Porshnev, Alexander Mashkovtsev, heard rumors from the Abkhazia region of the Caucasus Mountains of a wild woman captured by peasants. Knowing of Porshnev's interest in such stories, Mashkovtsev passed the tale along to him. If correct, the story of Zana would be a powerful element in Porshnev's argument. Porshnev traveled to the remote region and interviewed a number of individuals who claimed firsthand knowledge of Zana. Given many different details and dates—none of which had any collaborating records—Porshnev synthesized the story as best he could. Born sometime in the mid-nineteenth century and dying in 1880, Zana may have been a female Almasti. Captured and removed from her natural world, Zana became a curiosity. Subsequently sold or passed around, she came into the possession of a local nobleman named Genada, who took her to his estate near the town of Tkhina. Her life at the estate consisted of one degradation after another, including being kept outside in a series of enclosures and often shackled. Peasants threw rocks at her and generally abused her. Thought more animal than human, with an apelike face, she never learned to speak. Her bestiality apparently did not stop some men, possibly even the nobleman Genada, from knowing her intimately. Zana gave birth numerous times over the course of her sad life of captivity. Some of these children survived. A daughter named Gamassa lived into her sixties, and her youngest, a boy named Khwit, passed away as late as 1954. Reports Porshnev collected claimed the entire family possessed extraordinary physical strength, but with dark, ruddy complexions and simian facial features did not score high marks for beauty. In 1964 Porshnev tried unsuccessfully to find Zana's and Khwit's graves in order to examine their remains to check their biological status. Also a supporter of the relic theory, Myra Shackley said Zana and her offspring showed "the links between Almas and Neanderthal man in this instance seem quite strong."[66] An entire family of genuine Almasti lay buried frustratingly just out of Porshnev's reach. If only they could be found, he could prove the relic theory once and for all and show the skeptics how right he had been.

The tale of Zana is not the only such Almasti story. There have been a number of popular legends handed down about Almasti encounters, including a number that took place during World War II. One of the more popular tells of a Lt. Colonel Karapetian whose men captured what they thought to be a Nazi infiltrator. Apparently knowing something of paleoanthropology, the colonel believed his men had captured a live Neanderthal. After the war Karapetian tried to find out what happened to the strange captive, but all traces of the creature had disappeared.[67] Russian peasant folklore has a number of Almasti sightings and encounter stories that researchers point to as proof, just as Western investigators hold up the eyewitness accounts of Sasquatch.

Boris Porshnev did not take up the cause of wild men alone. As he unsuccessfully searched the Pamir Mountains for a Snowman, Czech anthropologist Emanuel Vlček inadvertently found one in Mongolia. In 1958 Vlček (1925–2006) took part in an expedition to Mongolia sponsored by the Czechoslovak Academy of Science. The expedition members sought to determine if the conditions there favored a larger anthropological study of the Mongolian Khalka people to see if any connection existed between them and the native people of North America. Vlček concentrated on literary materials in local libraries and archives in order to get a clearer picture of Khalka society and cultural practices. While looking through the library of the former lamaistic school in the town of Gandan, he uncovered a curious tome: "I found a book," he said, "by Lovsan-Yondan and Tsend-Otcher, entitled in free translation *Anatomical Dictionary for Recognizing Various Diseases*."[68] An herbal and natural history, the book fit the type he hoped would give insights into early Mongolian society. To his amazement, as he leafed through it he came across a page of illustrations showing on the right side a group of monkeys and on the left an image of a wild man. The pictures had captions in Tibetan, Chinese, and Mongolian, which Vlček ascertained "denote this creature in translation as man-animal."[69] The illustration was a fairly simple, yet realistic, line drawing of a manlike monster, faintly smiling and standing on a rock waving to the viewer. Chinese printers had for years been reproducing Mongolian texts in so-called "Peking Editions." The copy Vlček had come across had been printed in Peking in the late eighteenth century.

While going through the library of the Scientific Committee of Mongolia, Vlček came across another copy of the *Anatomical Dictionary*, this one printed in the nineteenth century in the Mongolian capitol of Urga (which had been renamed Ulan Bator— Red City—by the Communists). The captions in the book said the creatures had their origins in the bear family, yet resembled man. Vlček thought the creature might be the Almas then being studied by Mongolian linguist academician Rinčen.[70]

The illustration of a wild man, or Almas, discovered by Emanuel Vlček in Chinese and Mongolian texts has since been reproduced innumerable times in articles, books, and television documentaries on manlike monsters, and is often held up as textural evidence for the Yeti. However, according to Vlček the earlier of the two editions— the Peking Edition—is dated only back to the eighteenth century.

Likely it was a Chinese copy of an earlier Mongolian or Tibetan book, but no such earlier copy has been located. Vlček, an anthropologist and linguist of some ability, believed the *Anatomical Dictionary* suggested the creature was real. He cited the accurate descriptions and illustrations of monkeys in the same book and the overall realistic approach to medicine and natural history the book displays. He compared this book to typical medieval European bestiaries and medical books which regularly mixed real and mythical animals on its pages, and noted that the *Anatomical Dictionary* did not engage in such juxtapositions. If the creature shown in the *Anatomical Dictionary* did represent a mythical being, it was the only example in the heavily illustrated tome. Therefore, Vlček reasoned, it must represent a known species.[71] Although intrigued by his find, Vlček never mentioned any connection to relic Neanderthals.

### Dmitri Bayanov

Boris Porshnev's Neanderthal theory for the origins of the Almasti did not find wide support in mainstream Russian science. His championing of Marxist evolutionary theory did not keep his monster work from putting him on the outs with the Politburo. He never received official censure which, as in the cases of Baradiin and Žamcarano, could have proven fatal, but after the Pamir expedition he never again received official support and found it difficult to get his writings on the subject published by the state press. He did find supporters and followers, however. One of his closest was Marie-Jeanne Koffmann, born in France, but having lived most of her life in Russia. Koffmann held a medical degree and was an accomplished surgeon and had fought with distinction at the battle for Moscow, during WWII, leading a unit of soldiers. After the war, she became interested in the Almasti and did research into the topic as well as taking part in the 1958 Pamir expedition. She claimed the creatures "could be an ancestor of man."[72] She still pursued the topic as late as 1992, when she took part in a joint French-Russian venture back to the Pamirs, sponsored by the Russian Society of Cryptozoology.[73] Organizers were excited that public support came from Yves Coppens, the highly regarded French paleoanthropologist codiscoverer, along with Donald Johanson, of the *Australopithecus Africanus* fossil commonly called "Lucy." Koffmann remains an avid believer of the reality of the Almasti, supports the relic theory, and holds an important and revered place in the world of Russian monster hunters.

Another of Porshnev's champions, Dmitri Bayanov, promoted the relic idea. Born in 1932 and raised in Moscow, Bayanov and his family moved to the relative quiet of Tajikistan as the Nazi invasion steam-rolled toward them in the 1940s. There, he first heard stories of the Almasti, became fascinated by nature, and began to dream of the life of a zoologist. Even distant from the front lines of the war, his family suffered terrible privations, living at the edge of starvation. After the war he went back to Moscow and attended college but did not take any advanced degrees because "my scientific interests and works have been in subjects [relic hominoids] ignored or rejected by mainstream science."[74]

First meeting Boris Porshnev in 1964, Dmitri Bayanov quickly immersed himself in monster studies. He adopted Porshnev's views on relic Neanderthals and his rejection of being called a monster hunter. Like Grover Krantz in America, Porshnev and other Russian investigators did not think they chased monsters, and they certainly did not engage in the paranormal. They were scholars pursuing an unusual, but straightforward, scientific mystery about an undocu-mented species of animal. To this end, in 1970 Bayanov coined the term *hominology* to describe the search for Almasti and other such creatures. The Almasti represented an early ancestor to humans, and were not bogeymen. He later suggested the term *crypto-anthropology.* Bayanov argued that Porshnev deserved credit for making homi-nology acceptable because only he had the nerve to do it. With a bit of cold war rivalry lingering, Bayanov took Western authors to task over monsters and evolutionary theory. In his book on Bigfoot, John Napier, Bayanov said, had resisted rewriting evolutionary his-tory, while Porshnev tackled it head-on. "I dare say," Bayanov added, "Napier's book is written without much regard for truth....I guess the author was a sincere and sort of involuntary skeptic."[75]

When, after the failed Pamir expedition, the Soviet Academy of Sciences closed the Snowman Commission and ended their support of monster studies in 1958, Russia was left with no academic body look-ing into the topic. To remedy this, Porshnev's sympathetic colleague, biologist Pyotr Smolin (1897–1975) inaugurated a lecture series in 1960 at the Darwin Museum in Moscow. Both men thought some-thing as innocuous as a scientific seminar would attract the atten-tion of other academics, but not too much official state attention. Held monthly, and dubbed the "Relic Hominoid Problem Seminar," it allowed local enthusiasts, both amateurs and academics, to meet to discuss news of the field, hear lectures, and generally communicate. When Smolin died, leadership of the series passed to Dmitri Bayanov, who renamed the project the "Smolin Seminar on Questions of Hominology." In 1997 a special meeting was held there to celebrate the thirtieth anniversary of the Patterson film. Grover Krantz and John Green attended. The seminar is now in its fiftieth year.[76]

In the 1970s Bayanov's colleague in hominology, Igor Bourtsev, did something their mentor Boris Porshnev had been unable to do: he found the grave site of Zana's supposed son, Khwit. He then discovered Zana's skull, had the remains examined in Moscow, and the reports proved promising. The DNA of Khwit seemed consistent with that of a Neanderthal. In 2007 the National Geographic television channel, as part of their "Is it Real?" series, aired an episode titled "Russian Bigfoot." They recounted the Zana story and sponsored a new genetic test which seemed to show that Khwit, as well as the skull of his mother Zana, showed no genetic structure inconsistent with a modern Homo sapiens. Khwit was not the offspring of a union between a human and a relic Neanderthal; therefore Zana could not have been an Almasti. The findings did not please Bourtsev, Bayanov, or anyone who maintained the Zana story was genuine.

Weighing in on Khwit's apparent mundane origins, American anomalist writer Lloyd Pye took National Geographic to task. Pye gained prominence through his promotion of human remains purported to have been born of a human mother and extraterrestrial father. Pye called it the "Star Child." Supposedly found in a cave in the Mexican state of Chihuahua in 1930, the Star Child—a deformed child's skull—came into the possession of Pye, who promoted it as the remains of a human-alien hybrid. Genetic tests showed the DNA to be consistent with known Native American peoples. Critics argue the strangely inflated skull is of a four or five year old child who lived about 900 years ago and who suffered from progeria, the so-called "aging disease," from hydrocephaly, or some other genetic disorder. It may also have been a case of cranial binding known to have been practiced in Mexican culture. Despite the claims of its otherworldly pedigree, independent tests showed the Star Child's genetic origins to be completely human. As someone whose unusual evidence had suffered at the hands of "experts," Pye spoke out about the treatment of Khwit.

After explaining how the National Geographic's DNA tests of Khwit's skull misled viewers as "blatant distortions of fact," Pye hinted at darker machinations. Still smarting from the reaction the scientific community had given his Star Child, Pye accused National Geographic of covering up any information that would undermine their position as purveyors of scientific wisdom. "The National Geographic

Society," he said in a tone straight out of Ivan Sanderson, "is perhaps THE [his capitols] bastion of conservative mainstream thought in the world today, so they will do or say *anything* [his italics] necessary to protect their bailiwick."[77] Just what National Geographic's "bailiwick" consisted of, Pye did not elaborate on. Science had undermined not just Zana and her progeny, but spacemen. There would be more than DNA testing of ABSMs causing East and West to clash over monsters.

## The Trouble with Troglodytes

Dmitri Bayanov had great respect for Americans and America. Relief supplies from the United States had saved him, his family, and many of his neighbors from starvation during the war. He saw evidence of anomalous primates in North America as support for the Neanderthal relic theory in Russia. He traveled to the United States once and consulted with many of the manlike monster enthusiasts there, considering them friends.[78] He played a leading role in the formation of the International Society for Cryptozoology (ISC) along with Grover Krantz and others. Yet, he did not completely agree with how his American and Canadian colleagues went about their business.

Along with Boris Porshnev, Bayanov originally subscribed to the position that relic hominids should not be counted genuine humans. Along with Porshnev, he believed the only real unique marker of humanness was speech, and ABSMs did not have it. By the twenty-first century, however, Bayanov had changed his mind about the muteness of the Almasti, based in part on the Carter Farm Case from Tennessee.[79] Janice Carter Coy claimed her family, beginning in the 1970s, had been interacting with living Sasquatch-like creatures for half a century. Chronicled in *50 Years with Bigfoot* (n.d.), by Janice Carter Coy and Mary Alayne Green, the Carter family lived alongside these creatures and, especially interesting from Bayanov's point of view, had learned to converse with them in their own language as well as in limited English.[80] Igor Bourtsev went to the scene to investigate and was impressed by the story and the local people. He thought he recognized visual signs left by the creatures as signals to each other about human habitations and activity.[81] This seemed to confirm that relic hominoids not only still persisted, but stood much closer to being human, if not actually so, than he and Bayanov ever thought possible. Most in the Sasquatch research community saw the Carter Farm Case as dubious, particularly the claims of tool use and preferences for human food, made in the story. This reluctance disappointed Bayanov. Here was evidence of a population of these creatures living among humans, and no one seemed to care. He lamented that "mainstream science is turning its back on hominology."[82] Bayanov considered the Almasti semi-human troglodytes rather than apes. That Loren Coleman and others called the creatures "apes" also upset Bayanov. He told Grover Krantz that Coleman's work "is simply harmful to our credibility."[83] Along with the Carter Farm Case Bayanov always felt American monster researchers never gave the Patterson film the respect it deserved.

In December of 1971, René Dahinden brought the Patterson film to Moscow to show researchers there. To Bayanov, Dahinden seemed "like Santa Claus descending from heaven upon Moscow, carrying with him a sack full of wondrous gifts."[84] As they had in New York two years before, the monster enthusiasts hastily set up screenings of the film. They showed it to the staff at the newspaper *Izvestia*, at the Darwin Museum, the Museum of Zoology at Moscow State University, the Institute of Anthropology, and others. Just as in New York the reactions came mixed at best. The Russian scientists said the creature in the film "should not be hairy, that they can't have hair-covered breasts, all the usual things that American and Canadian 'specialists' had been saying." Even Boris Porshnev had his doubts.[85] After seeing the film, Bayanov thought it closed the case on the question once and for all. He felt the North American monster enthusiasts did not pursue the issues with enough vigor. He complained to International Society for Cryptozoology secretary Richard Greenwell that the group should be doing more to fund research into the Patterson film. Greenwell told Krantz, another founding member, "I am afraid we have not been on the same wavelength [with Bayanov about how cryptozoology should be pursued] since the society's founding." Bayanov complained that Greenwell and others should do more to get academics to view the Patterson film. Greenwell certainly agreed that more scholars should see it, but he told Krantz "we can't force scientists to study the film."[86] Bayanov blasted Greenwell, saying, "I am sure when the history of the ISC is written this will stand as the worst failure of the society."[87] Like so many monster enthusiasts Dmitri Bayanov found himself frustrated with what he considered solid evidence not being thought of as such by others. Why did they not see what he saw? Grover Krantz asked the same question.

## Conclusion

It can be argued that the greatest problem dogging cryptozoology has not been the gathering of evidence so much as the presentation of that evidence. How an artifact or a finding is put forward to the scientific community is just as important as the artifact or finding itself. From his student days, Grover Krantz had a quirky way of presenting his findings, casting doubt on their veracity. He never could quite overcome it. In 1979 he submitted a paper on Indo-European peoples in Anatolia to a journal. The editors rejected it because Krantz put forward too many undocumented assumptions, he did not know the current literature on the subject, and his knowledge of historical linguistics did not show enough depth. While the editors acknowledged that scholars "need dissenters [like Krantz] to keep the inquiry moving" they faulted him because "the author's snippiness gets in the way."[88] A paper he sent to *Evolutionary Anthropology* in 1997 was rejected for similar reasons. Rather than reading like a straightforward scholarly paper, Krantz's text came off as an opinion piece with no citations. "Krantz offers a conspiracy theory," the reviewer said, "that biological anthropologists are scheming to dismiss any valence to racial categories as a means of dissembling social racism." Such a theory, the reviewer concluded, "is absurd."[89]

The world's monster hunters had been working hard to present their findings, convince their colleagues, and legitimize their work. They experienced only nominal success. More academics had become interested in this work, but no major breakthrough had occurred. Every time a new discovery came out, expectations were heightened, but then crumbled. Every track print, every Minnesota Iceman, every photograph or film, every logical-sounding theory just bounced off the walls of mainstream science. Hoaxes did not help the cause either. The number of high profile cases dismissed by academics and critics as hoaxes only made it harder for the next finding to get a hearing. Now, at the end of the twentieth century the Founding Fathers of monster hunting had reached the end of the line. They had done all they could do, but what did they have to show for it?

# Chapter 7

# A Life with Monsters

*Having lost this battle almost totally, I am reluctant...to pursue this line any further.*[1]

Grover Krantz

In 1998, as he sat in his office for the last time before he retired, surrounded by plaster casts of Sasquatch footprints and hominid skulls that filled wooden shelves to the ceiling, Grover Krantz pondered his long career. Over the years he had collected hundreds of different animal skulls and the better part of a dozen human skeletons. All of them seemed stuffed into this room like a Renaissance cabinet of curiosities. On a table facing him the life-sized reconstruction of a *Gigantopithecus* skull stared at him with empty, but searching eye sockets, as had his dog Clyde's skull 25 years before. Had it all been worth it? One of Krantz's great complaints centered on the accusation that his career in the anthropology department of Washington State University suffered neglect and ridicule because of his work on Sasquatch. Had battling the conventional wisdom of academic scientists and the quirky nature of some of the amateur naturalists provided any tangible results other than disappointment and bitterness? Had someone undermined his career from the outside, or had he contributed to it himself? Had he resolved anything in the battle between the crackpots and the eggheads? Had it been worth living a life with monsters?

Unlike Ivan Sanderson, Bernard Heuvelmans, and other amateurs who could engage with monsters and complain about academe without the worry of retaliation—you cannot risk a job you do not have, nor ever will—Krantz lived in the community of professional scholarship. That community had protocols, rules for presenting evidence, and ways of behaving that its members expected. He had colleagues and supervisors upon whom his job, his reputation, and his security depended. While Krantz could give in to exaggeration and hyperbole, his complaints about conspiracies behind his back and dark whispers in the ivory halls were not without substance.

## Washington State University

In 1968 Krantz left the University of Minnesota teaching fellowship and took a position on the faculty of Washington State University in the anthropology department. A year after he arrived on the Pullman, Washington, campus of WSU, the Cripplefoot incident occurred. At the time Krantz had a general interest in cryptozoology and anomalous primates, but he had seen little evidence to support serious investigation. Like any junior faculty member in the first year of his appointment, Krantz searched to find a topic on which to focus his research, and he searched for his academic self. He had many interests, but had yet to find an intellectual cause. Bossburg gave him his epiphany. Years later, he recalled that "before I examined these [Cripplefoot] prints I would have given you ten to one odds that the whole thing was a hoax."[2] When he arrived at Bossburg he saw tracks for the first time. Stepping away from John Green and bending down to look at the Cripplefoot tracks in the snow, Krantz found himself as much as he found Sasquatch. Upon his return to Pullman, word spread of his sojourn into the field. He alleges that as a result of his trip to Bossburg, certain "now retired" faculty members tried to have his appointment terminated. Undeterred, Krantz defiantly said that "if it weren't for Sasquatch, I would still be a maverick."[3] Carleton Coon agreed. In a review of one of Krantz's articles, Coon praised Krantz by stating that he "has never been one to leave his neck in when he could stick it out."[4] Summing up his different attitude towards Krantz's work, a local Washington newspaper commentator chortled condescendingly that he, too, was "deeply involved again this spring trying to prove the existence of the Easter Bunny and the Tooth Fairy, if I have the time."[5]

While his critics and some faculty attacked him, his students loved him. Krantz had a reputation for engaging students with thought-provoking questions, challenging them to think harder. They considered him brilliant and quirky in a charming way. When he ran into difficulty getting promoted, students banded together and sent petitions to the department chair and the dean, encouraging them to support him. As he had at Berkeley, Krantz cut a memorable figure on the WSU campus. Tall and gangly, he presented a sight to behold and students grew accustomed to his antics. He could be seen loping across campus trying to imitate the stride of Bigfoot, or showing up in the cafeteria wearing a protruding prosthetic brow ridge intended to give him a sense of how *Homo erectus* viewed the world, in what Christopher Stringer called "an experiment of delicious eccentricity."[6]

In the 1970s, when Krantz first began to be recognized for his position on Bigfoot, Robert Ackerman, acting chair of the WSU anthropology department from 1971 until 1972, said of him, "we're all for this sort of research."[7] Krantz did receive funding from WSU for his mainstream research, but not for the monster work. Years later, Ackerman remarked that while the department did not support Krantz's Sasquatch research, it did not openly oppose it either. "We ignored his dalliance with Sasquatch," he said, and insists resistance to Krantz's promotion came, not because of Bigfoot, but because he did not publish quickly enough at first. "Grover was highly regarded as a scholar," Ackerman continued, "though many of us were amused at his persistence in pursuing things Sasquatch." As to Krantz's more acceptable work on hominids he added that Krantz "had some real insights on *Australopithecus* and felt that some of the reconstructions were possibly not correct, but did not want to take on the established leaders in the field."[8] Krantz told Bernard Heuvelmans that he received no help or encouragement from his colleagues, that the press misquotes him, and that his work on Sasquatch is either denigrated or simply ignored.[9] Heuvelmans commiserated with him over the "attitude of stuffed shirt scientists who simply reject the whole problem" of anomalous primates.[10] A later department chair, Geoffrey Gamble, admitted that Krantz's monster research "probably has had a retarding effect on his career." Reiterating Gamble's contention, Krantz added that "It's like being a woman or a Black; you have to do things far better in order to get accepted."[11] Krantz said bitterly that the school supported his Sasquatch work "only to the extent that I have not been fired."[12] WSU spokesman Al Ruddy commented in an almost apologetic tone, as if aimed at parents who might be nervous about what their children learned with Dr. Krantz, that "those who study with him say he's an excellent teacher and doesn't talk about Sasquatch in his classes."[13]

Krantz's promotion record at WSU is unusual in some aspects, though not all. He began at WSU as an assistant professor 1968 without a doctorate.[14] Krantz came to WSU as part of a hiring streak brought on by a National Science Foundation grant to build an interdisciplinary Quaternary Age studies program to study the geology and paleontology of the most recent period of the Earth's history. According to WSU documents, his first promotion, in September of 1972, raised him to associate professor, occurring the year after he had received his doctorate from the University of Minnesota. In 1974 he was awarded tenure. A university professor in the American public higher education system normally receives tenure prior to their first promotion from the assistant to associate level (tenure is not considered a promotion). Most state institutions like WSU have a waiting and evaluation period of five to seven years before tenure would be awarded (private universities can be different). The unusual order in Krantz's case shows some department members refused at first to vote for his tenure. Despite this, receiving his doctorate made his promotion easy, but he still needed to fight to get tenure, which he eventually did. Despite the apparent resistance of other faculty members, receiving tenure and promotion within six years of beginning at WSU is within the normal range and cannot be cited as a significant issue in his career progress.[15] Krantz viewed it differently, however.

The way Krantz saw it, the anthropology department "suffered from dissention and demoralization" over the years, mostly from decisions about tenure and promotion. Already convinced that hiring and retention came from personal issues rather than the proper consideration of teaching evaluations and publications, Krantz spoke out. Having, himself, finally made the hurdle to tenure and promotion, he could now comment on the issue. He saw what he considered poor decision making, with faculty members being retained when others who had performed as well or better were not reappointed or promoted. To address the issue, Krantz wrote to Dean Allan H. Smith, the man who had personally opened the anthropology department in 1950. He told Smith of his concerns and acknowledged he stood alone in making the complaint, telling the dean he would be unable to find any fellow faculty members who would corroborate his point because "they all fear for their own futures."[16] Smith, who openly expressed contempt for monster hunters, assured Krantz no inappropriate administrative judgments had been made, that all such decisions about tenure and promotion came after careful consideration of only academic concerns, and that "we believe them sound and appropriate."[17] Krantz tried again a year later to bring this issue out over another tenure denial, but received the same reply. Chastened now, Krantz turned his back on collegiality at WSU and focused inward on his own work and students. This turn away from engagement with his department pushed him further from his colleagues but closer to monsters and monster hunters, and contributed to the price he paid over his next run for promotion. The strained relationship between Krantz and the department was obvious. One former student from the WSU microbiology department of the 1990s—who prefers to remain anonymous—remembers the staff roughly divided into two camps when it came to Dr. Krantz and his promotion to full professor. Some liked or tolerated him, others thought of him as an embarrassment. This student said the faculty interacted with him in the hallways as if he were a crazy uncle, while students saw Krantz as an odd, but deeply devoted scientist who loved teaching and instilled a love of learning in his own students.

This situation did impact Krantz's promotion to full professor, a wait normally in the ten-to-twelve-year range after promotion to associate. His application was turned down several times. The promotion came for Krantz only through the efforts of then department chair, John Bodley, who had joined the anthropology department in 1970, two years after Krantz. A cultural anthropologist studying both southern and northern Native American groups, Bodley took to Krantz and they became friends. While not sharing Krantz's intensity over Sasquatch, Bodley did feel enough evidence existed to warrant a closer look into the subject. As many did, he considered Krantz's mainstream work quite good, but never really understood his monster obsession. When Bodley became department chair in 1992, he set about getting Krantz promoted. (Bodley himself eventually rose to the level of regents professor). He had his work cut out, as there were still several faculty members embarrassed by Krantz's work and who were resistant to rewarding him. Bodley managed to get a number of outside recommendations as well as student petitions of support. Finally, the faculty voted, and in 1994 Krantz became a full professor, 22 years after his arrival on campus.[18] Had it not been for John Bodley's efforts, Krantz would have remained an associate professor the rest of his career.

## To Shoot or Not to Shoot

If Krantz's study of Bigfoot generated controversy within the academic community, his ideas on how to study the creature did the same, and even more so, in amateur circles. In 1971, realizing he would have to produce a Sasquatch carcass or bone fragments, Krantz placed a small item in the local newspaper, the *East Washingtonian*. He called on anyone who might have shot a Sasquatch or run one over (accidentally) to contact him. He promised readers "anonymity can be assured."[19] Since childhood, he felt comfortable recovering the skeletons of dead animals. Recovering a Sasquatch skeleton in this manner was no different from any of the mice or cats he had dissected in his youth, or the primates or humans he had dissected in his career. From the start Krantz openly called for a specimen to be obtained, if not as a road kill then by purposefully shooting one. He understood, in ways many of the Bigfoot aficionados did not, that as distasteful as shooting one of the creatures would be, allowing them to go extinct for lack of proper wildlife management, or simple recognition, stood as an even more unpleasant alternative. While not a field biologist, Krantz did go driving through the woods on occasion armed with a rifle so that if he encountered a Sasquatch—he never did—he could acquire it himself the hard way.[20]

Krantz's statements that a Sasquatch should be shot did not go unnoticed. When a housewife read an article in which Krantz reiterated the need for a specimen to be acquired through this method, she sent him a copy of the news item with a note in the margins saying "couldn't life be saved and remain an intriguing mystery?"[21] Krantz received so much mail over his shooting comments he began sending out a form letter reply. He said his reason for making such statements was to ultimately protect the creature. He told concerned correspondents, "Everyone will be standing around wringing their hands saying: if only we knew they were real we could have saved them." He continued, "Well, they could have been saved if only we would blow one away now."[22] The form letter finished by reminding readers that "those few of us who study this animal on our own time and money are ridiculed by the rest of the scientific community."[23]

As experts and perceived authorities, scientists receive questions and requests for advice from individuals and the media about their specialty, especially if it is a controversial topic. In his role as Bigfoot 'expert' Grover Krantz engaged in correspondence not only with the active elite Sasquatch researchers like John Green, René Dahinden, and Bernard Heuvelmans, but with members of the general public. Some inquired as to the nature of these creatures, some recounted tales of meeting the creature in the wild, and some wrote exuberantly over the thrill and romance of monster hunting.[24]

Rather than viewing monstrosities and animal monsters with fear and trepidation, twentieth century people eagerly engaged with them. To meet this need, American, British, and European men's adventure magazines from the 1940s through the 1970s such as *Argosy*, *True, Je Sais Tout* and newspapers like *Illustrated London News* ran a steady stream of monster related articles eagerly consumed by a wide readership.[25] While some of the serious monster hunters like René Dahinden often cursed Krantz and generally disparaged his qualifications and position as a professional mainstream scientist, the less serious but still interested enthusiasts sent Krantz correspondence that was generally praiseworthy, supportive, and respectful of his position as a university academic.[26]

The idea of Bigfoot attracted children as well as adults. An example of this came during the mid-1970s when a clutch of young monster enthusiasts formed a club in Vermont. They had seen Krantz on television and immediately contacted him to let him know they stood alongside him in the search. They accepted Krantz's authority without question and took it as a given that the creature existed. Kirsten Francis, leader of the Vermont group, wrote to Krantz to say, "My friends are studying about Bigfoot." She claimed to have seen a Bigfoot and included a drawing of what she had seen.[27] Kirsten and her friends Cheri, Anne, and Stefanie took their work seriously. In a follow-up letter they told Krantz, "We have a laboratory in our basement. We meet everyday.... Our laboratory is not very much but it is enough." The girls even put together a fairly sophisticated theory to account for the monster. "Bigfoot," they said, "is a mixture of A. Rubustus [sic] and early Homo sapiens."[28] Krantz did not respond to their letters quickly enough because Anne Marie wrote separately to let him know they were not just kids fooling around. Being kindred souls she said, "We believe in Bigfoot because we have witnessed the shock of bigfoots passby [sic]." They had tried to take a picture or make a cast of the footprint, but the snowy ground resisted. She then waxed philosophic: "We hope that someday we do find the answer to all are [sic] monsters." She signed the letter "youre admire[er]" and added a little smiley face.[29]

Amateur enthusiasts commonly saw Krantz as a scientist who went out on "expeditions" to capture or kill Sasquatch, and they sought advice for their own proposed field operations. As enthusiasts wrote the American Museum of Natural History for help, a number also wrote him, wanting to join his excursions. In July of 1977 Californian Robert Kaup wrote asking advice for mounting an expedition. Kaup—who referred to himself as "Crazy" Kaup—had been reading up on Bigfoot and now wanted to go find one. "I am thinking on looking for the Bigfoot," he told Krantz, "as soon as possible (money wise)." He assembled a list of equipment he thought he would need, including guns, cameras, "knifes" [sic], and a tent.

He asked Krantz if he had left anything out, and if he knew of any expeditions going out that year. Krantz also received offers of help. A Pennsylvania correspondent claimed a friend had seen Bigfoot in the woods of Oregon. Luckily, his mother had an insurance settlement coming to her and he planned to send Krantz $25,000 to finance an expedition. "Please believe me," he went on, "when I say if I help the scientific establishment by proving its [Bigfoot] existence that's all I want." A few days later the budding philanthropist dejectedly told Krantz he was not going to be able to give him the money after all: his mother did not share his enthusiasm for giving away a large chunk of her settlement.[30] Krantz, in fact, never went on any formal expeditions and actually ventured into the field on a limited number of occasions, never seeing the creature and only seeing footprints *in situ* once.

Correspondents also had news of historic events. One of the more popular Sasquatch stories involved "Jacko."[31] In 1884 near Yale, British Columbia, a railroad crew came upon an ape-like creature lying injured near the right-of-way. They brought it to the sheriff's office and locked it up. Word spread of the incident and people began arriving at the jail to see the creature, dubbed Jacko. Before the crowd could see the creature, the railroad workers claimed Jacko had escaped. Today, the Jacko story is generally dismissed by researchers as a journalistic prank, but those who believed the story wondered what had become of Jacko. In 1970 Krantz received a letter from Canadian Chilco Choate, who claimed to be the grandson of the Yale railroad engineer present at Jacko's capture. Choate told Krantz that his grandfather related how he helped put the hapless, but still living, Jacko on a boat to London. The creature died in transit and had "simply been thrown overboard."[32] When little bits like this came into his office, Krantz dutifully filed them away for use when needed. He never gave correspondents money and rarely any advice for hunting monsters. This stemmed partly from his worry that his work already had a less than stellar reputation at WSU, and people from the lunatic fringe using his and WSU's name would only make things worse. In addition, part of Krantz's underlying agenda was to oust such amateur enthusiasts from the field to make it professional. Enough of these people already stomped about the field; he did not want to encourage the arrival of more. Grover Krantz stood at the crossroads of monster hunting, where the interested public, elite amateur naturalists, and scientists came together. He tried to negotiate his way through it all to retain and enhance his academic reputation, but the lines demarcating these groups did not hold firm.

## Serious Leisure

The search for Bigfoot and other anomalous primates can be seen as much as a search for legitimacy and status as a search for hairy, bipedal monsters. This held for the amateurs as much as for the academics. The academics wanted to be able to study these creatures without ridicule, and the amateurs wanted acknowledgement that what they did should be counted as legitimate as anything done by an academic with a degree and a university posting. In the case of science practitioners of the latter twentieth century, one way to distinguish professionals and academics (from amateurs) could be to group them as those who earn a substantial part of their living in the scientific field, who have academic degrees, or have some type of organizational affiliation with a government office, corporate body, museum, or institution of higher learning where their job requires them to perform the functions of the field. Did such positions exist in monster hunting? Arguments are made that even amateur status cannot exist unless those amateurs have at least a theoretical chance to become professionals. For example, "one cannot be an amateur butterfly catcher or match collector; no opportunity for full-time employment exists."[33] In other words, amateur status does not exist if there is no chance to become a professional or if there are no academics to be compared with. In the case of monster hunting, could such a transition from amateur to professional occur? Monster hunting—at least the search for manlike monsters—did have the potential for becoming academic. This idea motivated many of the scientists involved. They wanted to professionalize and legitimize the field and displace the amateur leadership. If Bigfoot or the Yeti did exist, they would be counted as primates. Studying them would fall under the purview of the established mainstream fields of anthropology and primatology. Therefore, an amateur Sasquatch enthusiast could go out and acquire the requisite training to become an academic Sasquatch researcher.

None of the scientists discussed here could be counted full-time researchers of cryptozoology. By the end of the twentieth century, Idaho State University's Jeffrey Meldrum, successor to Grover Krantz's position by investigating anomalous primates, came close—even suffering many of the problems Krantz did in the form of ridicule from some of his peers and the media.[34] Scientists who did monster work, with the exception of Krantz's few scholarly articles in the 1970s and those by Meldrum some years later, stepped out of their place and entered the domain of the amateurs, as amateurs engaged in a leisure time hobby—albeit with more intellectual training and technical skills. No professorships in anomalous primate studies or cryptozoology could be acquired. There have been scattered classes offered at the college level, but mostly as histories of the field rather than as training for professional scientific work.[35] If the definition of professional status includes earning a living from the work one does, or having an affiliation with a museum or university, then it has been some of the amateurs who more closely fit the definition of professional monster hunter rather than their academic counterparts. Ivan Sanderson made his living writing on cryptozoological and other esoteric topics, and Bernard Heuvelmans had a science doctorate and museum affiliation. Peter Byrne found private and even occasionally public funding to support him in his monster work. John Green earned money from his books, while René Dahinden made up a good bit of his meager income from the rights to the Patterson film. Calling these individuals professionals, however, still posed problems. Thomas Kuhn described science as a community effort. Scientific communities have shared ideals, values, and goals. The monster hunters could be seen as a loose community, but they did not share many ideals, values, and goals, other than the desire to prove the creatures exist. Complicating this issue, many of the amateurs cared little about being ranked next to their academic counterparts. They tended to give themselves their own legitimacy and felt no need to try to become part of a mainstream system they had mixed feelings about.

Most monster hunters, whether academic or amateur, performed their cryptozoological activities on what could be termed leisure time. As Nathan Reingold argues in his analysis of professional and amateur science, Robert Stebbins asserts in his study of leisure time pursuits that dedication to the pursuit outweighs any monetary concerns. Under what he calls "serious leisure," Stebbins identifies a number of prerequisites for counting someone as a serious amateur, a status the monster hunters achieved. They have long-term commitment to what they are doing, have developed skills associated with the activity—use of technology such as night vision and video recorders, tracking skills, historical textural research, and others—and have built their lives around this activity rather than around the work they do for pay. Far from seeing what they did as a mere hobby, the elite cryptozoologist/monster hunters considered what they did as anything but trivial. It took center stage as the driving passion of their lives. For serious monster hunters their activity was more than just a good time, an enjoyable day or two in the woods. What they did, they argued, had the possibility of contributing to the furtherance of science.

The monster hunters do fall somewhat short in one important area Stebbins points out. As a group the monster hunters constructed few value sets, shared ideals, or ways of thinking about the activity that are uniform across the discipline.[36] Regardless of what they are called or what values they did or did not share, the amateur naturalist monster hunters held their work so closely it would be difficult to let it go should the creatures ever be found.

Even if manlike monsters were proven genuine biological realities and became the realm of mainstream science, the amateur practitioners would not throw up their hands and leave the field. It would likely prove impossible for amateurs to gain access to any Sasquatch remains once turned over to a university, however. This would keep them on the outside, unless they found remains of their own. They would be ousted as leaders in the field, but they would still pursue their work, the pronouncements of anthropologists or primatologists notwithstanding. Before they could take over, however, the academic scientists involved in looking for cryptids needed to establish a community structure for themselves.

## The ISC

Continuing with his project to legitimize the field of monster hunting and bring it into the realm of anthropology, in 1978 Grover Krantz helped organize the first scholarly conference on the subject. His anthropologist partner in the endeavor, Dr. Marjorie Halpin (1937–2000) of the University of British Columbia, studied Native American Tsimshian society and did important work on totem poles of the Pacific Northwest coastal cultures. The conference took place on the campus of UBC, where years before René Dahinden and John Green had their run-in with faculty scientists over the Patterson film. The conference, the first to bring together a group of academics to discuss anomalous primates, worked so well that its success gave sanction to think more along this line might be done. In 1987 Krantz and Halpin collaborated on the book *Manlike Monsters on Trial: Early Records and Modern Evidence* as a collection of papers given at the conference.[37]

Plaguing the search for manlike monsters had always been lack of search techniques and unified effort. The Tom Slick and *Daily Mail* expeditions worked at being systematic, but as Carleton Coon and George Agogino found, various internal arrangements made that difficult. Things that hold mainstream scientific communities together include shared and uniform methodologies, common goals, and a system by which information can be shared. The search methods of the monster hunters had been at best haphazard, their collation, presentation, and sharing of data virtually nonexistent. Like alchemists, enthusiasts in the early decades of cryptozoology hoarded and guarded their information rather than share it. Those who understood the importance of sharing work had few venues to do so other than one-on-one exchanges. This frustrated scientists Coon, Agogino, and Krantz, and even some amateurs as well. By the 1970s the number of scientists interested in the anomalous primate phenomena began to grow, but they still tended to keep a low profile. Heartened by the success of the 1978 UBC conference, a number of researchers banded together to do one of the other traditional things needed to organize a field and its practitioners. In the early 1980s an organization formed to help codify ideas and unite scattered researchers under a respectable scientific banner.

Discussions about such an organization between the University of Chicago's Roy Mackal, the Smithsonian Institution's George Zug, and University of Arizona Office of Arid Land Studies coordinator Richard Greenwell, resulted in the formation of the International Society of Cryptozoology (ISC) at the Smithsonian Institution in September of 1982. It proved easy to interest scientists from universities and museums around the world in becoming officers. The board of directors included Leigh Van Valen (University of Chicago), Dmitri Bayanov (Darwin Museum, Moscow), Zhou Guo Xing (Beijing Natural History Museum), Phillip Tobias (University of Witwatersrand, South Africa), as well as Grover Krantz. Membership included the likes of Dr. Christine Janus, Brown University paleontologist, and others of equally significant rank and reputation. Greenwell held the position of society secretary and chief cheerleader the rest of his life, with Bernard Heuvelmans as its first president.[38]

Immediately upon founding the ISC, members worked to explain their goals, philosophy, and operative methods. In the first issue of the society's journal, *Cryptozoology*, Heuvelmans gave his description of what the field entailed.[39] He likened it to paleontology, which to his mind searched for unknown animals from the past. Heuvelmans desperately wanted to convince readers that cryptozoologists undertook real science worthy of respect. He began by saying cryptozoology did not involve any arcane, supernatural, or occult practice. Reconstructing cryptids was no more unusual or suspect than when paleontologists reconstruct fossil creatures. Cryptozoologists engaged in a process of "demythifying [sic] the content of information in an attempt to help make the inventory of the planet's fauna as complete as possible."[40] They looked for animals yet to be described scientifically. Heuvelmans also attempted to explain *how* those aims would be achieved. He argued that the method by which this happened started with a reference or description of some scientifically unknown animal taken from folkloric sources, either written or oral tradition. Then researchers looked to known "representatives" similar to the folkloric accounts. This allowed the researcher to "sketch the salient features of its behavior, to determine the nature of its habitat [and to] approximate limits of its area of distribution." This process in turn would allow the researcher to go to find, attract, and collect the creature. The ISC's primary function, though, was to give scientists an outlet for the research they did on cryptozoological issues, and which they could not get published elsewhere. The society's membership grew quickly. Academics at both small colleges and at universities of the Ivy League joined. Articles in the journal ranged from those on the perception of the Loch Ness Monster to Pygmy Gorillas, the extinction of thylacines in Australia, Russian cultural artifacts depicting extinct animals, as well as field reports from around the world. Annual conferences were held across North America and England. Cryptozoology seemed alive and well as respected scientists began to step forward with their research.

The 1980s proved a busy time for Grover Krantz. Along with helping found the ISC, he visited China in June of 1982 to discuss Wildman research going on at Beijing's Natural History Museum; he met with fellow ISC board member Zhou Guo Xing and also visited Dmitri Bayanov in Moscow. In 1989 Krantz and WSU hosted the annual meeting of the ISC. Noted amateur naturalist researchers Daniel Perez and John Green gave presentations. Terry Cullen also spoke. The now Dr. Cullen had been the man to first alert Ivan Sanderson to the presence of the Minnesota Iceman, the act which kicked off the incident. He still believed real the creature Colonel Hanson exhibited and that Sanderson and Heuvelmans never had a chance to get a good look at it the way Cullen had.[41]

Krantz's mind was continuously in action, addressing various issues. He hoped to establish a weekly commentary column in the local *Daily Evergreen*, where he could let his stream of consciousness run free. After seeing the movie "JFK," he hurriedly scribbled down, "Tibbets was assigned to get Oswald." A liberal pro-choice advocate, he supported abortion on demand and chided any male anti-abortion supporter, saying only women would ever have to face such an ordeal so only the woman herself should be allowed to decide. His desire for an outlet for his curmudgeonly musings never quite materialized.[42]

While his monster work often brought derision, Krantz's mainstream anthropological work generated respect. He gave over a hundred invited talks to academic and popular groups and acted as a consultant to local police and sheriff's departments on reconstruction of shoes and skeletal parts, occasionally giving courtroom testimony. In 1981 he examined evidence of a man claiming to be Charles Lindbergh's son, showing he was not. He traveled widely, making trips to China and Russia to research human evolutionary questions. In the early 1990s, he made several trips to Java to study *Homo erectus* (Java Man), and other fossils, and to retrieve casts to bring back to Washington. He chaired the panels of at least nine doctoral and 11 master degree students. He consulted on the Kennewick Man case and argued the nine-thousand-year-old skeleton could not be traced to local, modern native peoples and, therefore, should be kept by scientists for study. The lead investigator of Kennewick Man, James Chatters, had been one of Krantz's graduate students and brought him in on the case. This skeleton, found by hikers along the Columbia River in Washington in 1996, immediately caused a controversy because of its unusual age.[43] Local Native Americans argued the skeleton was one of their ancestors and as such the recently passed Native American Graves Protection and Repatriation Act required it be returned to them for burial. Scientists claimed the skeleton too old and too morphologically different to be a native ancestor, so it should not be repatriated, but studied. Krantz's work supporting the cause for study rather than burial drew the ire of the Umatilla tribe, partially because of his history of monster work and partly because they accused him of "using outdated techniques" and "nineteenth century science" to come to his conclusion about the remains.[44] The Umatilla thought it inappropriate and insulting to have a cryptozoologist deciding the status of their ancestor.

Despite his active participation in the ISC, Grover Krantz never openly articulated a cryptozoological philosophy, as Heuvelmans had. This may have stemmed from his belief that, despite what the media might think, he was an anthropologist focusing on unusual primates, not chasing monsters as a cryptozoologist—a term he rarely used in public or print. His goals included bringing anomalous primate studies to anthropology, not anthropology to anomalous primate studies. Bernard Heuvelmans and Richard Greenwell took a separatist stance, eager to create a new discipline anchored in an established tradition. Krantz and most scientists interested in the topic stood as reformers seeking to add a new chapter to the mainstream. For all its good intentions and work, the ISC had a relatively brief life. In the early years of the twenty-first century, the group's two great moving forces, Heuvelmans and Greenwell, passed away. With no one interested in taking their places, the organization went extinct. The International Society of Cryptozoology stood as a watershed, paving the way for other similar groups in the United States, United Kingdom, Russia, France, and Sweden. The amateur monster hunting impulse still ran hot, and growing numbers of individuals and organized groups banded together and kept up the search. The individual cold warriors and ideologues, mountaineers, and rugged outdoorsmen of the 1950s and 1960s were replaced by groups that included a mixed bag of both the Left and Right politically oriented, environmentalists, New Age spiritualists, and others, all with energy and drive. They in turn were joined by a growing number of academics, laboratory researchers, and field biologists. The era of René Dahinden making his long solitary treks through the forest primeval examining his soul; carrying nothing but a backpack, a rifle, and a light heart faded into history.

### The New Monster Hunters

Loren Coleman considered himself a professional cryptozoologist. By 2010 he had been pursuing monsters for half a century. He earned all of his living from his monster related work: he went into the field; he became the face of cryptozoology, appearing in the media as a paid consultant on such matters, and between 1989 and 2004 taught classes on the subject at the University of Southern Maine. He published his research prodigiously, though not in peer reviewed journals.[45] He had come to know many of the golden age monster hunters, chronicling their work for newer generations of enthusiasts. In 2009 he opened the International Cryptozoology Museum—a first—in Portland, Maine. As Green, Dahinden, and Krantz became models in the past, Coleman now was a new model. His position allowed him to act as a linking element between the old and new generations.

By the turn of the century a new school of individual cryptozoological investigator had emerged, first seen in the career of Loren Coleman. Interest in cryptozoology and manlike monsters had steadily grown in England ever since initial reports of the Yeti had attracted worldwide attention. In 1992 English cryptozoologist Jonathan Downes established the Centre for Fortean Zoology in Devon as a way to focus interest on cryptozoology through its journal, *Animals and Men*. He also began a publishing house, Fortean Zoology Press, to produce works on the topic. A number of other English researchers came to prominence in the 1990s, including Richard Freeman, who became increasingly fascinated by cryptids while studying zoology at Leeds University. Freeman joined the Centre for Fortean Zoology and began writing and researching.[46] He traveled to numerous countries around the world searching for a wide range of cryptids. Nick Redfern followed in Ivan Sanderson's tracks, writing magazine articles and books.[47] Zoologist Karl Shuker, who had a doctoral degree in animal physiology, made his living as a cryptozoological researcher and writer, regularly going into the field. Unlike Americans such as Jeffrey Meldrum, who focused on anomalous primates, the English investigators tended to be generalists whose work extended over a wide range of species. These researchers found themselves joined by the next trend: the monster hunting group. The elite researchers could muster the means to travel widely in their search for cryptids. The groups tended to be local affairs, not venturing far from home, and searching only for cryptids thought to inhabit their neighborhoods. These groups were more insular than the globetrotting elites, but still searched with great passion.

Though the International Society of Cryptozoology had welcomed amateur naturalist members and allowed them to speak at conferences, the majority of the published work in its journal came from professional scientists. Following in the wake of the ISC, enthusiasts began to form their own clubs. In 1991 the International Bigfoot Society formed in Oregon. Members met regularly to discuss the latest sightings, controversies, and other topics and publish a newsletter. In 1995, what is likely the largest and most influential group, the Bigfoot Research Organization (BFRO), formed, calling itself "the only scientific research organization exploring the Bigfoot/Sasquatch mystery." The membership included amateurs and academics alike.[48] The BFRO helped pioneer the trend in collecting data from sightings and other research efforts and publishing this data on their website for wider consumption. It encouraged members and nonmembers to send in collected material for posting. The BFRO has been followed by a number of other amateur-oriented Bigfoot societies, such as the Southern Oregon Bigfoot Society and Texas Bigfoot Research Conservancy, which sponsors a popular yearly conference and get-together.[49] There are groups now located in most of the regions of North America where Sasquatch is regularly sighted, as well as in a number of European countries.

For nonscientists, Bigfoot was tailor-made for aspiring cryptozoologists. The primary requirement for entering the hunt was simply a desire to do so. Depending on where one lived, enthusiasts wanting to take part in the adventure need only walk out their back doors to enter the world's largest anomalous primate laboratory. Unlike the amateur naturalists of the eighteenth and nineteenth centuries, who saw themselves as erudite, sophisticated virtuosi, the North American Sasquatch community consisted mostly of working-class and lower middle-class white males in their thirties and forties with outdoor and hunting experience—or at least an understanding of such environments—with modest or no college education. Of the monster enthusiasts with university training, only a few studied fields such as geology, anthropology, or environmental sciences. Not concerned with former cold war issues of East-West relations, the new monster hunters worried about deforestation and the encroachment of urbanization on wild places, about native peoples issues, endangered species, Global Warming, and the metaphysics of man's relationship with the natural world. Many became involved because of prior personal experience encountering the creatures. The largest political affiliation is with the Democratic Party, twice as many as Republicans, followed by those with no affiliation, then Libertarians, then Independents. The religious preference is Christian, with only a sprinkling of atheists, a lesser number of pagans, and few if any Jews, Moslems, or other non-Christian denominations. While most subscribe to a belief in some form of evolution, a surprising number identify themselves as creationists or theistic evolutionists, who believe in a divine presence which uses evolution or divine intervention to account for living species and the changes they show in the fossil record. It is not uncommon to hear a response such as, "I don't think evolution has all the answers," to questions about natural selection accounting for the presence of anomalous primates. The overwhelming majority of Bigfoot research organization members support the idea that Bigfoot is an elusive, but straightforward, biological creature whose existence can be proven once and for all with perseverance and a little luck. About 10 percent think it is an "inter-dimensional being," with another 10 percent thinking it is an alien, nonterrestrial entity. Finally, the majority feel that half the mainstream scientific community believes them, while the other half does not. As of the early twenty-first century, only a few still think that all scientists ridicule them.[50]

The profile of a rural class of Bigfoot hunter should not be seen as unusual, given that anomalous primates have rarely been seen strolling down Michigan Avenue, Detroit, or Park Avenue, New York. Many of the characters in the history of Bigfoot hunting—whether as researcher, witness or hoaxer—shared this background. They were people comfortable in the wild, appreciative of its grandeurs and respectful of its dangers, whether the dense woods of the Pacific Northwest, the plains of Texas, the swamps and bayou of Louisiana, or the forests of New York State. They did their work on a part-time basis, having another livelihood that allowed them to pursue their passion when time allowed. Only a small percentage of Bigfoot hunters earned a living from the field. Those who did worked mostly as writers on the subject or hired themselves out as trackers, tour guides, or lecturers and, increasingly, with the proliferation of cable and satellite television programs on unusual phenomena, as on-screen hosts and talking heads.

Far from being backwoods bumpkins or simpletons, however, what the amateur groups lacked in formal training and scientific acumen they made up for in enthusiasm and drive. They quickly made ample use of the Internet to construct a cyber college of the mind, exchanging ideas, engaging in discussions and arguments, and passing along information. Roger Patterson's 16mm camera gave way to sophisticated gadgetry like video and DVR cameras, audio recorders, motion detectors, and other high tech accoutrements. Collected biological materials could now be sent off for DNA and other laboratory tests that Carleton Coon and George Agogino did not have access to. The new generation still reveled in their amateur status, and like their forebears still strove to be as "scientific" as possible. They communicated with, and befriended, their heroes like John Green, René Dahinden, and Grover Krantz. The original generation of Sasquatch hunters had now reached a standing as wise old sages, scarred veterans of the rough and tumble early years. Younger enthusiasts looked to them for inspiration and validation.

## The Old Guard

The new generation may have been coming to the fore, but the old guard still had things to say and do and argue over. René Dahinden's relationship with certain members of the Bigfoot community continued to be fiery, and he still reserved special wrath for Grover Krantz. So much commotion came from it that ISC secretary Richard Greenwell referred to the Krantz/Dahinden relationship as "all out war."[51] A naturally talented and intellectual man, Dahinden's upbringing kept him from the type of formal education and career he would have excelled in. That he spent a good part of his adult life eking out a meager living from the back of a trailer in a gun club parking lot could not help but instill resentment. He sat in his trailer on cold Canadian nights, woodstove burning, listening to classical music, smoking his pipe, and musing on various subjects like a Cambridge don in a well appointed office dispensing wisdom to younger seekers, or just being alone. To Dahinden his disappointment, frustration, and anger at mainstream science found focus in the person of Grover Krantz. After their initial meeting at Bossburg, their relationship went through periods of calm, even friendliness and cooperation, but no matter how Krantz tried to mollify or placate Dahinden the smallest spark could fire up the Swiss-Canadian's passions.

Dahinden seemed to fight with everyone. Just days before Christmas of 1975 he had an encounter with Peter Byrne in the parking lot of a McDonald's restaurant in Portland, Oregon. The Irish adventurer and Swiss-Canadian had never been close. As they argued about comments Byrne made in the infamous *True* magazine article, Byrne turned to leave. The discussion had begun earlier at another location, and Dahinden, intent on keeping it going, had followed Byrne to the burger joint. As Byrne walked away Dahinden chased after him, yelling obscenities. Byrne turned to face the pursuing Dahinden and punched him. The police soon arrived, and tempers settled. Neither man wanted to press charges, so after taking some details the police left. The two went their separate ways.[52]

Having little else than his Sasquatch materials, Dahinden reacted strongly whenever he thought someone had invaded his space or looked as if they might take what he considered his. Over the years he had acquired a considerable collection of first-generation plaster casts of footprints, but he did not have the only collection. John Green had an impressive collection as well. Grover Krantz had been sent a number of casts over the years and thought he might make a few dollars off of them. Krantz put together a mail order business to sell plaster Bigfoot casts and stills from the Patterson film. As a state university employee, Krantz's bosses at WSU wanted to ensure neither he, nor by extension the university, could get in any legal entanglements over the plan. After looking into the subject, the Washington State attorney general said "the rights with regard to the tracks are more difficult to define with legal certainty" than the film stills. He continued: "It is my opinion that an authentic track would not be copyrighted since such a track is not an original work of authorship." He suggested, however, that WSU and Krantz should "proceed with caution" and stay within the fair-use doctrine and avoid sale of the stills altogether.[53] Krantz eventually worked out a deal with a Kennewick, Washington, store to sell the tracks, while his former student, Ira Walters, sold them by mail order. The WSU anthropology museum sold casts for a time. Dahinden wrote to Richard Greenwell, worried about Krantz's selling plaster copies of Sasquatch footprints. Dahinden feared a market flooded with copies undermined his collection. "I mean," he asked Greenwell, "if everybody has them—what do I have left???"[54] Dahinden had taught himself passionately, but imperfectly, as a naturalist and had turned his entire life over to Bigfoot research to the point where he had little else to sustain him. To lose that would have been to lose his life. While Dahinden complained of scientists' reticence to investigate Bigfoot, he was wary of them when they did look at the phenomena. In 1985 the *Seattle Post Intelligencer* said Dahinden "knows that when Sasquatch is proven to exist, the scientists will move in and take over and [he] is dead. He'll be shoved aside."[55] This made Krantz a useful target for Dahinden's sharp tongue and uncompromising demeanor. He badgered Krantz about the casts, claiming they were his so Krantz should ship them all to him.[56] Krantz's years of osteological preparation and casting experience resulted in his footprint cast work being of a high order. Giving that away to Dahinden just because he demanded it seemed ludicrous to Krantz. "Just once in your life—LISTEN," the enraged Dahinden screamed in his note to Krantz, after the scientist said he had no intention of giving him anything.[57]

Dahinden guarded his rights to the stills from the Patterson film with special tenacity, hunting down breaches of his position. In addition to the original, a number of good quality copies of the film rested in the possession of John Green, Peter Byrne, Grover Krantz ,and Jon-Eric Beckjord. The ever-animated Beckjord tried to sell his copy in online auctions.[58] Having a copy of the film worried Krantz. He told Dmitri Bayanov, "Rene does not know I have located a good copy of the film to study, and when he finds out he will do everything he can to stop me from using it in any way."[59] When Krantz's *Big Footprints: a Scientific Inquiry into the Reality of Sasquatch* (1992) appeared, it included drawings made from stills from the film but not the stills themselves. In 1999 Dave Hancock of Hancock House Press reissued the work as *Bigfoot/Sasquatch Evidence*. In this edition Krantz printed two small stills from the Patterson film. Initially, Krantz also wanted to use a picture of Dahinden holding casts, but in the end did not. As he considered all these pictures his property, Dahinden again demanded royalties, but Krantz refused to pay. Krantz had begun working on the manuscript for the book after the Bossburg incident and had told Dahinden about it. When Dahinden objected to Krantz's using the stills, he was puzzled. Krantz thought he had an "understanding" with Dahinden over using the images, royalty free. Dahinden, however, had no intention of having any understanding that did not bring him an income from the film. Krantz told the hot-tempered monster hunter that he thought of him as a friend—at that moment anyway—and "would like to continue this arrangement if you will let me."[60] Dahinden would not let him, but Krantz thought he could still mollify Dahinden. In 1975 he congratulated Dahinden on his successful attempt to gain property rights over the Patterson film.[61] Despite Krantz's optimism, others thought he was being naive in dealing with Dahinden. "If Grover thinks René is not upset with him, Grover is mistaken," John Green confided to another. "René has been upset with Grover for a long time."[62] After reading the infamous *True* magazine article, Dahinden became infuriated because Krantz had said complimentary things about Byrne, whom he disliked. He sputtered to Krantz, "every time you open your mouth to the press you make a bunch of down right stupid statements." He went on, "Patterson called you an opportunist years ago, and I guess he was right." He then threatened Krantz, "I will pull you down and blackball you in the Sasquatch research." He signed the letter, "yours truly."[63] Even John Green, who had a good relationship with Krantz, was a bit upset over the *True* article. Green never told Krantz anything about Byrne in the way the article suggested. Green had warned Krantz that he should be wary of him. Dahinden disliked Byrne and said of him, "It is a question of dealing with known liars and conman [sic] and make believe artists."[64] Peter Byrne himself thought the article of "poor quality" and many of its assertions "pure fabrication" and referred to the author, Al Stump, as an "odd fellow." Despite this, Byrne told Krantz "I do appreciate the kind remarks you [made] about me and my work."[65] Upset with what was happening, Krantz wrote to *True* magazine, claiming the article had libeled him. The magazine replied that they had not. Krantz told *True* representative Robert Gottlieb that he felt he deserved some monetary compensation because the article and people's reactions to it "have already damaged my professional reputation and exposed me to public ridicule by my peers."[66] The magazine refused to accept any responsibility, but agreed to run a rebuttal written by Krantz, which appeared in the October issue. He corrected what he felt were the many misquotations and misrepresentations the article contained especially those regarding his *Gigantopithecus* theory.[67]

Still battling over ownership rights into the mid-1990s Dahinden considered any copies of the Patterson film an infringement on his royalties and demanded they be handed over to him. As Krantz had anticipated, Dahinden flew into a temper over his having a copy. Dahinden argued Krantz must pay a portion of the royalties for his book to him because he had used film stills to illustrate it, or else be sued. Writing to WSU faculty member John Bodley, Dahinden said, "After reading his chapter on the Patterson Film in his crazy book, more than ever—I demanded that he give back the film." Dahinden then declared, "That chapter is so dumb and stupid that it boggles the mind."[68] When a positive review of the book appeared by 23-year-old monster enthusiast Mike Quast, Dahinden wrote the reviewer, berating him for giving a positive review to a work he thought trash.[69] Dahinden also hounded Richard Greenwell over photo issues. Exasperated, Greenwell wrote back telling the Swiss bulldog not to write him anymore. He just could not take the aggravation. "I will pay you the same attention," Greenwell told Dahinden, "as I pay to Strasenburgh, Beckjord, Perez, etc.,"—others Greenwell thought poorly of.[70]

Dahinden wanted to sue Krantz over the use of the film stills. In a letter to a Canadian law firm concerning legal action, Krantz called Dahinden a "nut case" after Dahinden left a "condescending and offensive voice mail" on the firm's answering machine.[71] The tornado around Dahinden dragged in anyone unlucky enough to step too close. When John Green remarked on the situation of the films and ownership, Dahinden turned his sights on him. Green operated the newspaper in his hometown of Harrison Hot Springs, British Columbia, sometimes called Harrison Village. In one of his angry letters to Krantz, Dahinden referred to Green, a university trained journalist, as "the Harrison Village Idiot."[72] Right up to the end of his life, Dahinden threatened lawsuits. In addition to Krantz, the list of potential litigants included Peter Byrne, Richard Greenwell, and Hancock House, which had published Krantz's book. Dahinden wanted $500 from the publisher for use of the contentious stills.[73] It was all bluster, though. The question of ownership of the film and stills was such a muddle that no one had a clear idea of the legal status. Jon-Eric Beckjord argued the film was in the public domain and therefore could be used by anyone royalty free. The attorney general of Washington State told Grover Krantz that René Dahinden's claim of rights could easily be challenged.[74] Dahinden could have lost what rights he had, or thought he had, by actually going to court. His threats to sue were intended to get people to pay him without a protracted legal battle.

It was not all bad blood between the scientist and the naturalist, however. Krantz and Dahinden had years of relative calm, even partnership after a fashion. In the early 1980s Dahinden took a friendly stance towards Krantz because of his worry about the growing presence

of Russian investigators in North American anomalous primate studies. Dahinden traveled to Moscow in 1971 to show the Patterson film
to a collection of academics, who gave him a warm reception. By now
wary of the Russians, Dahinden warned Krantz to watch himself in
his dealings with the Soviet scientists he thought had hidden agen-
das.[75] Around the same time, Krantz and Dahinden came together
in their dislike of Jon-Eric Beckjord. Early on Krantz thought the
rambunctious Beckjord "a nice guy" and welcomed Beckjord's photography skills and manic energy. If this energy could be contained
and guided, Krantz told Dahinden, "We may all benefit."[76] The feeling was not mutual. In 1978 Beckjord wrote to WSU President Gene
Terrell accusing Krantz of unethical behavior and demanded he be
fired. Krantz said things about Beckjord he considered disparaging, which had helped ruin a writing deal with *National Geographic
Magazine.* Terrell handed the complaint over to WSU Dean B. A.
Nugent for investigation. Nugent could find no evidence that Krantz
had done any of the things Beckjord accused him of and, in fact,
found Beckjord's arrangement with *National Geographic* still held,
although it had suffered—not from anything Krantz allegedly said—
but from Beckjord's own tardiness in getting the article in on dead-
line. Therefore, Nugent told Beckjord, their professor's reputation
remained intact and that Beckjord should desist from making false
allegations in the future.[77]

When more shenanigans arose from Beckjord in 1987, Krantz
asked Dahinden—who also disliked his counterpart in agitation—
"Did it ever occur to you that Beckjord is in the pay of Weyerhauser or
some other big company to ridicule the Sasquatch hunt indirectly?"[78]
The giant logging company had interests in huge swaths of land in
the regions most often cited as a Sasquatch habitat. If Bigfoot turned
out to be real, it would be rated an endangered species and its home
might conceivably be made off limits to the logging industry. Whether
or not Dahinden saw dark motives in Beckjord's behavior or pup-
pet strings being pulled did not matter. He came to dislike Beckjord
intensely. He told Krantz, "I am glad to see that you too hate his
guts." He assured Krantz, "I will do my damdest [sic] to get rid of
him somehow."[79] Indeed, the world of monster hunting had many
strange bedfellows and much internal warfare.

While fighting with René Dahinden over the Patterson film,
Krantz had another Bigfoot sighting to be part of. In 1995 an avid
Sasquatch researcher named Cliff Crook purchased photos said to
be of a Bigfoot. Crook had previously worked as a technical advisor
on the Bigfoot themed film "Harry and the Henderson's" (1987).

A Forest Service worker eyewitness, whom Crook claimed wanted to remain anonymous, had come across a Bigfoot frolicking in a pond near Mount Baker in Snoqualimie National Park near Mt. Rainer, Washington. This individual managed to get to a bluff overlooking the pond and the creature as it waded about. Perilously close, the eyewitness snapped off a number of pictures, looking down on the creature just as it looked up into the camera. Crook purchased the pictures, which came to be known as the Willow Creek photos, for a reported $1,600, and took them straight to the tabloid newspaper, *The Sun*. The tabloid ran one of the pictures, along with commentary by Crook and Grover Krantz. Crook called the pictures the best evidence ever found of the reality of Bigfoot. Krantz said "There's no way to be sure, but they look pretty good[,]...that's all I can say."[80]

If genuine, the Cliff Crook pictures were astonishing. Unlike most monster photographs, shot at a distance and always blurry, the Willow Creek pictures came up-close, personal, and very clear. They are so close that they are almost portraits or publicity agency head shots. There is detail in the face as the creature stands dripping wet in the bright sun. The pictures, while taken from slightly different positions, show the creature in exactly the same position with no apparent movement between shots. Most reports of Bigfoot describe the creature as notoriously unwilling to hang about once seen. In the Patterson film, the moment the creature sees Patterson and Gimlin it stands up and walks off in a huff away from its pursuers. The figure in the Crook pictures just stares. This led skeptics to claim a hoax had been perpetrated using a manikin or model. The BFRO vigorously insisted Crook was a fabricator of hoaxes and a shameless self-promoter. Crook insisted with equal vigor that he was not.[81] As with all such sightings, indistinct pictures proved nothing and good, clear pictures proved to be fakes. As Krantz had shown guarded support for the pictures, critics argued the Willow Creek incident counted as just one more example of how easily he could be taken in.

### The End of the Golden Age

The turn of the century brought a thinning of the ranks of the old guard of monster hunters. René Dahinden, Bernard Heuvelmans, Richard Greenwell, and Grover Krantz all passed within a few years of each other. Marjorie Halpin had died in 2000; Dahinden and Heuvelmans followed in 2001; Ray Wallace, Jerry Crew's foreman, passed in 2002, the same year as Krantz. Immediately after Wallace's death, his family announced he had faked the famous Bigfoot prints Crew found. Richard Greenwell died in 2005, and Jon-Eric Beckjord in 2008. They all left indelible marks on cryptozoology, some more than others.

In the lore of monster hunting, René Dahinden stands as a tenacious and combative icon, yet his gruffness masked a lighter side. To those who came to know him well he was a good-natured, warmhearted man, as ready to have fun at his own expense as with others. He often philosophized about the loneliness of the hunt for monsters. At despondent moments he worried he might be chasing a ghost, but then accepted he had no alternative but to keep looking no matter what. In 1997 Dahinden received a diagnosis of cancer, the malady that hit the Sasquatch research community with special vengeance, also taking Greenwell and Beckjord. In his typically stoic way Dahinden shared the news of his illness with few people. One he did tell was longtime researcher and Sasquatch writer Christopher Murphy. Just before the end Dahinden confided to Murphy, "You know, I've spent over 40 years—and I didn't find it." With the resignation of one with no way left to turn, he said "I guess that's got to say something."[82] He died in April of 2001. Dahinden's voluminous papers went to his estate and fell into a kind of limbo when his trailer full of materials was moved to his son's farm, rarely being seen by other researchers or chroniclers of the world over which he cast such a long shadow.

In 1998 Grover Krantz retired from Washington State University, though he still held classes online. Thinking of the years of hard work, especially on the footprints and dermal ridges, he said "Having lost this battle almost totally, I am reluctant . . . to pursue this line any further."[83] His health began to deteriorate, but he kept writing and researching evolutionary mechanics. He continued to submit papers on non-Sasquatch issues for publication, but without success. His last attempt came in February of 2001, when he submitted "Neanderthal Continuity in View of Some Overlooked Data" to the journal *Current Anthropology*. Editor Benjamin Orlove rejected it. Krantz still suffered from a writing problem that had plagued him since his student days. His paper did not make use of, or even reference, the latest scholarship on the topic the way it should have. Whether consciously or not, Krantz's anger over years of abuse came out in his article. Orlove told him the tone of the paper was "jarring and unacademic."[84] Krantz rewrote and resubmitted the paper, but it was still unsuitable. In one last painful rejection, the blind reviewer said "the paper has an odd tone of someone who feels he has been wronged over the course of his career."[85] He decided not to resubmit the article. His career was over.

With his retirement Krantz moved to Port Angeles, Washington, with his wife Diane Horton. They met years before through their mutual interest in anomalous primates. In 2001 he received a diagnosis of pancreatic cancer. He suffered from digestive problems and increasing pain. At Washington's Peninsula Interdisciplinary Pain Clinic he received pain patches. His years of smoking contributed to his poor health. When WSU banned smoking in all its buildings Krantz, ever the rule breaker, set up a fan in his office near the window so he could continue to smoke and blow the evidence out the window so no one would notice. Ever the scientist, he searched online for information on his condition and what he could do about it. He grimly told Loren Coleman, "Medicine men differ as to whether I have months or years to live."[86] It was months. Just before the end, fellow WSU faculty member John Bodley visited one last time. For years Krantz claimed he had developed a foolproof method of determining a fake track from a genuine one just by looking at them. He steadfastly refused to tell anyone what his method entailed, so he could avoid hoaxers. Bodley, who since 1970 had been at WSU and one of Krantz's few open supporters on the faculty, decided to finally ask his old friend what the method was, insisting he would tell no one. Krantz just smiled and refused to answer.[87] He died the next day, taking many secrets with him.

## Conclusion

Cryptozoology, in a general way, has become more respectable in the biological sciences, but only just barely.[88] The amateur naturalists of the eighteenth and nineteenth centuries contributed in complex and subtle ways to the burgeoning of modern academic science. As the various fields of investigation grew in sophistication of thought and action, they generated data on geologic formations, botanical processes, bird populations, and animal behavior, contributing to biological systemization and evolution theory to guide and inform their work as well as that of future generations. The monster hunters of the twentieth century failed to ever contribute anything to zoology, primatology, or anthropology. Like their predecessors, they created a republic of letters and established societies, but only among themselves. Their ideas did not become part of the wider scientific discourse on primate evolution or behavior. The only theoretical structures made by the anomalous primate enthusiasts included Bernard Heuvelmans's

*Gigantopithecus* connection (1952), Ivan Sanderson's contention of ABSM diversity (1961), and Boris Porshnev's Neanderthal survival theory (1970s). None of these moved beyond their original incarnation or found common currency among their originator's intellectual descendants let alone mainstream science. While they tried to apply theoretical science and evolutionary thinking to monster hunting, Grover Krantz and the other academics involved suffered from a conceptual problem. Even scientists sympathetic to the idea of anomalous primates considered the evidence lacking. As far back a 1958 William Straus of Johns Hopkins University told the monster enthusiasts the reason he was unwilling to accept the reality of these creatures was because he was "only adhering to the basic tenet of scientific procedure when I ask for…positive proof of its reality."[89] The best evidence Krantz had was a theory. He extrapolated anatomical details from footprint casts and the Patterson film according to straightforward—if outdated—anthropological techniques. He rejected wild claims, accepting only empirical evidence he felt passed rigorous tests and guidelines for authenticity. He went about studying the creature in as sober, logical, and scientific a manner as he could. He did not fail completely, however. His work inspired other academics to join the hunt and keep it going. Despite little movement in the field, more scientists than ever now take seriously the idea of anomalous primates. Jeff Meldrum said "were it not for Grover, all this would have simply been dismissed."[90]

The interest shown by scientists over the decades could not quite shake some of the old problems. In 2010 noted English cryptozoologist Richard Freeman still believed strongly that mainstream zoology rejected cryptids because of nonscientific concerns. Echoing sentiments from a half century before, he said mainstream science "seems to be like some fundamentalist religion whose high priests say what can and cannot exist from the ivory towers of their comfortable lecture halls and labs."[91] They refused to look at the evidence before them and accept new ideas. Unlike the heroic narrative of scientists being either dismissive or antagonistic to the endeavor, however, a number of scientists not only have taken a sympathetic attitude toward cryptozoology, but have actively engaged in it as Freeman argued they should. These scientists were as excited as any amateur about the prospect of finding proof of the Yeti or Sasquatch. The problem did not stem from scientific disinterest. It stemmed from a lack of evidence. Scientists like Carleton Coon, John Napier, and Grover Krantz ran excitedly toward reports of proof, but time and again found themselves disappointed.

Scientists interested in the mystery-ape question looked everywhere. There seemed so many obvious and logical places to look. They looked at the roof of the world and across the mountains of Central Asia; they looked amongst the trees of Indonesia, California, Texas, Oklahoma, Louisiana, New York, and the forests of Staffordshire. They looked in caves, glades, jungles, and swamps, as well as in newspapers and books. They looked, but they rarely saw. One of the things that linked so many of the crackpots and the eggheads—John Green, René Dahinden, Bernard Heuvelmans, Ivan Sanderson, Peter Byrne, John Napier, William Charles Osman-Hill, Boris Porshnev, Carleton Coon, George Agogino, and Grover Krantz—was that none of them ever actually saw a Sasquatch or Yeti with their own eyes. They had assertions; they had logic, evolutionary theory, footprints, and hair samples, but nothing that proved beyond a doubt the creature's existence. The footprint casts were erratic and unpredictable in that they were not uniform. The significance of dermal ridges was recognized only by some investigators. The Patterson film was at best inconclusive.

It should be noted that Grover Krantz did not suffer because of his scientific radicalism. Rather, he was criticized for using out of date methods and for not being as well versed in genetics as he should have been. Krantz should not be seen as a martyr on the altar of unconventional thinking, although he thought of himself that way. None of his ideas or practices could be called revolutionary, or outside anthropological thinking. His approach was old fashioned rather than forward looking. His arguments swayed few colleagues, often leaving them unimpressed, if not baffled. Public statements that he could not be fooled by hoaxers undermined his status as a scientist and opened him up to ridicule when he was fooled.

Science is a community and a social endeavor. A discovery can be made by an individual, but the community validates the discovery and helps determine where in the scheme of things the discovery should be situated. Academics and amateurs have brought forward evidence they thought convincing, but the anomalous primate field did not make the changeover to accepted mainstream practice the way amateur ornithology, geology, and zoology had in the past. They preached to the choir, but the choir needs no convincing; the heathens and unbelievers are the ones who do.

In negotiating his career, Krantz found himself caught between the eggheads he wanted to convince and the crackpots he wanted to both exploit and distance himself from. The amateur naturalists did not want him taking their prize, and the professionals for the most part had little interest in what he brought them. He and other like-minded academics attempted to transform monster hunting into a legitimate scientific enterprise by applying evolutionary theory and accepted anthropological practice. He endeavored to legitimize his work and himself as a way of defining his place, and he found himself constantly thwarted in his efforts to convince others that what he did rated as genuine science.

In 1961, Ivan Sanderson said, with just a hint of glee, "I for one am looking forward with a good deal of pleasure to seeing what the 'experts' have to say when they come face-to-face with one of the thousands of 'Abominable Snowmen' which are living today...."[92] Around the same time, British primatologist William Charles Osman-Hill made an assessment of what he knew of manlike monster evidence and what he wanted to see. Osman-Hill had tracked anomalous primates in Southeast Asia, had been involved in the Tom Slick adventures in Nepal and California, and had viewed firsthand much of the physical evidence brought forward for the Yeti and Sasquatch. He thought it likely the creatures did exist. With all this as a background he said, somewhat disappointedly, about the Yeti that "it would seem that as far as physical evidence goes, apart from footprints, no absolute proof has been evoked for the existence of any animal in the Himalayan area unknown to science." He did not quite give up. He finished by saying hopefully that, despite this situation, "the data considered...suggest that something awaits explanation."[93]

Osman-Hill's colleague, George Agogino, also had been part of the hunt for manlike monsters from the beginning. Like Sanderson and Osman-Hill, he too believed the evidence suggested something was there. Unlike Sanderson, Agogino had always advocated waiting for sufficient evidence before grand statements were made or reports published. His advice mostly fell on deaf ears. He watched as legions of monster hunters made claims—some rational, some outrageous—on whether manlike monsters roamed the land. In 1978 he finally released a report of his own. Addressing the various forms of evidence put forward over the years he argued, like Osman-Hill, that most of it rated as little more than "distracting." Occasionally, something turned up that intrigued him. Though none of it solved the case conclusively, some of it "certainly keeps alive the possibility of its existence." His optimism drove him to claim that "I believe it is possible that we may obtain a Sasquatch specimen during the next decade."[94] If proof did not materialize by that time, he said he would give up on the animals being living creatures and support the folkloric nature of the phenomenon. His timetable came and went with no resolution to the problem and no apparent contribution to science.

In the closing days of the first decade of the twenty-first century, no monsters have been found. Back in Washington State, the snow still falls quietly on the pines around Bossburg and the other parts of Bigfoot's Pacific Northwest kingdom, as well as on the many far off places where difficult evidence lurks, and skulks, and peeps about. The ghosts of Sasquatch, René Dahinden, and Grover Krantz stalk each other in an eternal hunt for the unusual and out of place. Undeterred monster hunters, amateurs and academics alike—the crackpots and the eggheads—still creep through the brush, looking for them all.

# Notes on Sources and Monster Historiography

Those who searched for manlike monsters in the twentieth century—not as metaphors, but as flesh and blood organisms—have gone largely overlooked by academic historians of science. This field, as with cryptozoology in general, became the domain of independent amateur chroniclers producing a range of works of varying quality.[1] An excellent explanation of what cryptozoology attempts to do is found in Chad Arment's *Cryptozoology: Science and Speculation*.[2] Since the 1960s, scholarly works on anomalous primates, and cryptids in general, look to place them in the realm of legend and myth: creations of the human mind rather than of evolution.[3] These works tend to fall under what Jeffrey Cohen called "monster theory."[4] Works taking an empirical, physical anthropology approach include Gill, Meldrum, and Bindernagel.[5] Recent writings have begun to address the lives of the monster hunters, but follow the tradition of focusing on the folkloric and pop culture nature of Bigfoot rather than on the natural history element, and not on the place of cryptozoology in the context of the history of science. This category tends to lean to the exposé or dismissive side.[6] Of use to the discussion of monsters in general are scholarly works that attempt to put studies of human monsters into the history of biological systemization and classification.[7]

A number of methodological issues need to be addressed in the historiography of anomalous primate studies. There are papers collections of leading researchers. Grover Krantz, Bernard Heuvelmans, and Ivan Sanderson have accessible materials, as do Carleton Coon and George Agogino. The papers of other important scientists involved in the story—like John Napier and William Charles Osman-Hill—are harder to find, but are there in varying forms. The culture in which scientists are trained is one that promotes careful note taking and recording of work and the archiving of those records for later generations to utilize. When historians begin a research project one of the first things they do is identify where important papers are located so they can base their work on primary sources. On the amateur end of the spectrum, such papers collections are harder to find because the culture of archiving—donating their papers to a museum or library—has yet to make inroads into the monster hunter community. Daniel Perez has produced a useful bibliography of printed works on the subject.[8] The vast bulk of original source materials on North American monster hunters reside in private hands. Owners of private collections of monster hunter correspondence and primary documents can run the gamut of helpful to obstreperous, of easygoing to high-strung. The models of difficult monster papers collections are those of Tom Slick and René Dahinden. These two pivotal amateur researchers amassed large collections of documents, but their estates have been reluctant to let anyone, especially academic historians, have access to them. This means that an important part of the story will go untold or only appear as shadows at this point.

Locations of accessible papers collections are noted throughout the text. The largest and widest ranging collection is at the National Anthropological Archive of the Smithsonian Institution, Suitland, Maryland. This contains the papers of Grover Krantz, Carleton Coon, and George Agogino, along with scattered letters from many of the key monster hunters. Other Smithsonian archives contain correspondence pertaining to the Institution's role in the Minnesota Iceman case. UCL special collections, London, has monster related material from John Napier. The library of the British Museum of Natural History has a wonderful collection of now otherwise mostly lost British newspaper articles on anomalous primates. A portion of Ivan Sanderson's papers are in Philadelphia at the American Philosophical Society. Unfortunately, upon his death, Sanderson's papers were apparently looted, so the APS has only a portion of the original bulk. The APS collection is still highly useful, though. One of the more interesting collections of monster correspondence is in the Mammal Department Archive at the American Museum of Natural History in New York. As far as I can tell, this particular cache had gone undetected before I used it. After his passing, the bulk of Boris Porshnev's papers went to what was then called the Lenin Library, but what is now the Russian State Library, Moscow. His Almasti related materials are in the archive of the Russian Academy of Sciences. The Bernard Heuvelmans papers reside at the Cantonal Museum of Zoology in Lausanne, Switzerland. This collection represents a major reservoir of material on the history of monster hunting. There are undoubtedly more primary source materials still waiting to be discovered in various museum and university archives around the world as well as in all those private collections.

Less concrete, but still the cause of some problems, is the fact that as a group the monster hunters have no uniform research techniques or theoretical paradigms to guide their work. As a result each monster hunting group—and each individual monster hunter—must be approached separately. They resist classification into say, Darwinians, Neo-Lamarckians, or other evolutionary biology classifications or schools of thought. Many ideologies cross boundaries into the same organizations due to the fact that few monster enthusiast groups make rules of membership other than general interest in the topic, and even fewer approach their subject in the evolutionary way a biologist or primatologist would. This is great for democracy, but hell on historians. Some enthusiasts are evolutionists, while a surprising number are creationists. Many do not take evolutionary theory into account at all. Just when you think a category will work, it falls apart. There are, in fact, few points, techniques, or systematized thought that monster hunters agree on that the historian can use to organize their story. This book represents an attempt to work out and analyze the history of monster hunting that I hope others will follow.

# Chronology

1951    Shipton photos published; Boris Lissanevitch attempts a Yeti hunt
1952    Bernard Heuvelmans makes Yeti/*Gigantopithecus* connection in print
1953    Carleton Coon and Dillon Ripley make Yeti/*Gigantopithecus* connection in private; *Daily Mail* Expedition born; Edmund Hillary and Tensing Norgay climb Mt. Everest; Grover Kranz (GSK) begins to accumulate material on mystery-apes; René Dahinden arrives in Canada
1954    Carleton Coon makes Yeti/*Gigantopithecus* connection in print, Zana's son Kwit dies
1955    GSK graduates with a degree in anthropology; Bernard Heuvelmans, *Sur la Piste des Betes Ignorees*; Willey Ley uses term "Romantic Zoology," makes Yeti/*Gigantopithecus* connection
1956    Tom Slick goes to Nepal and meets Peter Byrne
1957    Harrison Hot Springs Expedition proposed; *Life* magazine expedition organized; Soviet media accuses Yeti hunters of spying
1958    *Daily Mail* Expedition to Nepal; Jerry Crew finds Bigfoot tracks at Bluff Creek; Bernard Heuvelmans, *On the Track of Unknown Animals*; A.G. Pronin sees Snowmen while on the Fedchenko Glacier; Soviet government forms the Snowman Commission and mounts an expedition to study them; Emanuel Vlček discovers Tibetan wild man in a book
1959    Slick expedition to Nepal; Peter Byrne switches bones in Pangboche Hand; Bud Ryerson finds Bigfoot tracks at Bluff Creek; Ivan Sanderson, "Strange Story of America's Snowman"
1960    Hillary-Perkins Expedition to Nepal; Vladimir Tchernezky reconstructs Yeti foot; Pyotr Smolin inaugurates the relic hominid seminar in Moscow; Academician Rinčhen creates finding aide to Žamcarano papers
1961    Ivan Sanderson, *Abominable Snowmen: Legends Come to Life*
1962    Tom Slick dies in plane crash
1964    Boris Porshnev and Dmitri Bayanov meet
1966    Roger Patterson, *Do Abominable Snowmen of America Really Exist?*
1967    Patterson-Gimlin film shot; John Green and René Dahinden visit Bluff Creek; Patterson-Gimlin Film screened at AMNH
1968    GSK begins at WSU; Heuvelmans credits Sanderson with coining term cryptozoology; Minnesota Iceman incident;

John Green, *On the Track of Sasquatch*; Ivan Sanderson, "First Photos of Bigfoot: California's Abominable Snowman"

1969  Bossburg incident; Ivan Sanderson, "The Missing Link"

1970  Dmitri Bayanov coins term hominology; GSK adopts *Gigantopithecus* theory; Roderick Sprague calls for articles on Sasquatch for NARN; John Bodley joins WSU faculty

1971  GSK publicly calls for shooting a Sasquatch; René Dahinden brings Patterson-Gimlin Film to Moscow

1972  GSK publishes his first scholarly article on Bigfoot; Roger Patterson dies, Boris Porshnev dies

1973  John Napier, *Bigfoot: The Yeti and Sasquatch in Myth and Reality*; Ivan Sanderson dies; René Dahinden, *Sasquatch*; GSK meets Cliff Crook

1974  Heuvelmans and Porshnev, *L'Homme de Neanderthal est Tujours Vivant*; Boris Porshnev's "The Troglodytidae and Hominidae" translated

1975  Peter Byrne, *The Search for Bigfoot: Monster, Myth or Man?*; Al Stump, "The Man Who Hunts Bigfoot"; René Dahinden acquires rights to Patterson-Gimlin Film; Scott and Rines, "Naming the Loch Ness Monster"

1978  George Agogino hopes Sasquatch will be proven by 1988; Man-Like Monster conference held at UBC

1979  GSK sees dermal ridges in footprint casts; Peter Byrne tells Sydney Anderson he knows Patterson-Gimlin Film is fake

1980  International Society of Cryptozoology formed

1982  Michael Dennett debunks dermal ridges; Mill Creek prints found

1983  John Wall coins term cryptid; Myra Shackley supports Neanderthal relic theory; Michael Heeney questions Baradiin sighting

1985  GSK, "A Species Named from Footprints"

1987  *Harry and the Hendersons*

1990  Ciohon, Olsen and James, *Other Origins*

1991  International Bigfoot Society formed in Oregon

1992  Center for Fortean Zoology established, Devon, England; John Bodley becomes chair of anthropology department at WSU

1995  BFRO formed; Bousfield and LeBlond, "An Account of *Cadborosaurus willisi*," Willow Creek photos

1997  Dmitri Bayanov, *America's Bigfoot: Fact Not Fiction*

1998  GSK retires

1999  GSK, *Bigfoot/Sasquatch Evidence*

2000    Margorie Halpin dies
2001    René Dahinden and Bernard Heuvelmans die
2002    GSK and Ray Wallace die
2003    *Homo floresiensis* discovered
2004    Dave Daegling, *Bigfoot Exposed*; Greg Long, *The Making of Bigfoot*
2005    Richard Greenwell dies
2007    Kwit's DNA determined to be completely human
2008    Jon-Eric Beckjord dies; Loren Coleman establishes Museum of Cryptozoology

# Notes

## Introduction   Chasing Monsters

1. René Dahinden and Don Hunter. *Sasquatch* (Toronto: McClelland and Stewart, 1973): 112.
2. Ivan Sanderson. *Abominable Snowmen: Legend Come to Life* (Philadelphia, PA.: Chilton, 1961).
3. Ibid., 446.
4. William L. Straus, Jr. "Myth, Obsession, Quarry?" *Science* ns 136:3512 (April 20, 1962): 252.
5. Eric Norman. *The Abominable Snowman* (New York: Award Books, 1969): 22. Eric Norman was one of the many pseudonyms of writer Brad Steiger who authored over a hundred books and articles on fantastic subjects.
6. Boyce Rensberger. "Is it Bigfoot, or Can it be Just a Hoax?" *New York Times* (June 30, 1976): 78.
7. Arthur Conan Doyle. "The Red Headed League," 1891.

## 1   Crackpots and Eggheads

1. C. W. R. D. Moseley, trans. *The Travels of Sir John Mandeville* (New York: Penguin Books, 1983).
2. This narrative comes from the recollections of Green, Dahinden, Krantz, and others.
3. See Al Stump. "The Man Who Hunts Bigfoot," *True* 56:456, May, 1975:28–31, 74–77. The quote is from Grover Sanders Krantz to Robert Gottlieb, 6/24/1975, folder 0334, box 3, Grover Krantz Papers Collection, National Anthropological Archive, Smithsonian Institution, hereafter NAA. Also from here Grover Sanders Krantz will be abbreviated as GSK.
4. Peter Byrne, *The Search for Bigfoot: Monster, Myth or Man?* (Washington, D.C.: Acropolis Books Ltd., 1975).
5. GSK, letter to the editor/rebuttal in *True* (October 1975):11.
6. Correspondence in GSK, box 7, folder 0334, NAA.
7. John Green to GSK, May 11, 1975, folder 0334, box 7, NAA.

8. René Dahinden to GSK, May 26, 1975, folder 0334, box 7, NAA.

9. Vladimir Markotic and Grover Krantz, eds. *The Sasquatch and Other Unknown Hominoids* (Calgary: Western Publishers, 1984): 147.

10. A poll conducted in the early 1980s found the majority of professional anthropologists in North America felt no acceptable evidence existed of Sasquatch and little justification for any funded research into it. Richard Greenwell and James E. King, "Attitudes of Physical Anthropologists towards Reports of Bigfoot and Nessie," *Current Anthropology* 22:1 (Feb., 1981):79–80.

11. René Dahinden to GSK, 5/12/1975, folder 0334, box 3, NAA.

12. Gian Quasar. Bermuda-triangle.org (2006).

13. Daniel Perez. Center for Bigfoot Studies, CA, "Review of *Bigfoot/Sasquatch Evidence*," 1999.

14. Jim Endersby. *Imperial Nature: Joseph Hooker and the Practices of Victorian Science* (Chicago: University of Chicago Press, 2008). The word scientist, first coined in the 1830s by the British philosopher William Wheuwell, did not see wide application until the later part of the nineteenth century and has been problematic ever since.

15. See Harriet Ritvo. *The Platypus and the Mermaid and other Figments of the Classifying Imagination* (Harvard University Press: Cambridge, MA, 1997), and Lorraine Daston and Katherine Park. *Wonders and the Order of Nature 1150–1750* (New York: Zone Books, 1998).

16. George Gaylord Simpson. "The Beginnings of Vertebrate Paleontology in North America," *Proceedings of the American Philosophical Society* 86 (1943): 130–88.

17. Robert Silverberg, *Mound Builders of Ancient America: The Archaeology of a Myth* (Greenwich, CT: New York Graphic Society, 1968) and Gordon R. Willey and Jeremy A. Sabloff, *A History of American Archaeology* (San Francisco: WH Freeman & Co., 1974).

18. A. Hunter Dupree. *Science in the Federal Government: A History of Policies and Activities to 1940* (Cambridge, MA: Belknap Press, 1957).

19. See, John C. Greene. *American Science in the Age of Jefferson* (Ames: Iowa State University Press, 1984); Brooke Hindle. The *Pursuit of Science in Revolutionary America, 1735–1789* (Chapel Hill: University of North Carolina Press, 1956); and Dirk Jan Struit. *Yankee Science in the Making: Science and Engineering in New England from Colonial Times to the Civil War* (New York: Dover Publications, 1991).

20. Mark V. Barrow, Jr. *A Passion for Birds: American Ornithology after Audubon* (Princeton: Princeton University Press, 1998).

21. Jeremy Vetter. "Cowboys, Scientists, and Fossils: The Field Site and Local Collaboration in the American West," *Isis* 99:2 (June 2008):273–303.

22. Brian Regal. *Henry Fairfield Osborn: Race and the Search for the Origins of Man* (London: Ashgate, 2002); and Ronald Rainger, *An*

*Agenda for Antiquity* (Tuscaloosa, AL: University of Alabama Press, 1991).
23. Robert E. Kohler. *All Creatures: Naturalists, Collectors, and Biodiversity, 1850–1950* (Princeton: Princeton University Press, 2006).
24. Nathan O. Hatch, ed. *The Professions in American History* (Notre Dame, IN.: University of Notre Dame Press, 1988).
25. Nathan Reingold. *Science, American Style* (New Brunswick, NJ: Rutgers Univ. Press, 1991).
26. Adrian Desmond. "Redefining the X Axis: "Professionals," "Amateurs" and the Making of Mid-Victorian Biology—A Progress Report," *Journal of the History of Biology* 34 (2001): 3–50.
27. Paul Lucier. "The Professional and the Scientist in Nineteenth-Century America," *Isis* 100:4 (December 2009): 699–732.
28. See Daston and Park. *Wonders and the Order of Nature.*
29. See Alixe Bovey. *Monsters and Grotesques in Medieval Manuscripts* (London: The British Library, 2002), and John Block Friedman. *The Monstrous Races in Medieval Art and Thought* (Cambridge, MA: Harvard University Press, 1981).
30. Zakiya Hanafi. *Monster in the Machine: Magic, Medicine, and the Marvelous in the Time of the Scientific Revolution* (Durham, NC: Duke University Press, 2000).
31. Thomas R. Williams. *Getting Organized: A history of amateur astronomy in the United States* (Doctoral Thesis, Rice University, 2000).
32. Peter Bowler. *Science for All: the Popularization of Science in Early Twentieth-Century Britain* (Chicago: University of Chicago Press, 2009).
33. Student records of Ivan Sanderson. Eton School registrar.
34. Student records of Ivan Sanderson, Trinity College Cambridge.
35. D. M. S. Watson. *Obituary Notices of Members of the Royal Society* 7:19 (Nov. 1950): 82–93.
36. *National Cyclopedia of American Biography* vol 57 (Clifton, NJ: James T. White & Co., 1977):192–94. Though anonymous this entry is commonly thought to have been written by Sanderson's second wife Sabina.
37. Department of Health, City of New York, summons to Alma Sanderson, March 1, 1951; Sanderson, Alma folder, Sanderson Papers, APS.
38. Ivan Sanderson. "There Could Be Dinosaurs," *Saturday Evening Post* (January 3, 1948).
39. Lucien Blancou. *Géographie Cynégétique du Monde* (Paris: Presses Universitaires de France, 1959).
40. Bernard Heuvelmans. "The Birth and Early History of Cryptozoology," *Cryptozoology* 3 (1984): 1–30.
41. Wall used the term in a letter to the editor of the newsletter of the International Society of Cryptozoology (ISC) in 1983 (vol. 2, no. 2, p. 10). He intended it as a way to refer to an individual animal that

fell under the purview of cryptozoology. It was picked up quickly and is still used extensively.

42. Bernard Heuvelmans. "What is Cryptozoology?" *Cryptozoology* 1 (Winter 1982):1–12.

43. Heuvelmans. "Birth and Early History."

44. "Obituary of Bernard Heuvelmans." *Fortean Times* 153 (December 2001).

45. Pierre Assouline. *Hergé: the Man Who Created Tin Tin* (Oxford: Oxford University Press, 2009): 170–172.

46. For the life of Huxley see, Adrian Desmond, *Huxley: From Devil's Disciple to Evolution's High Priest* (Reading, MA: Addison-Wesley, 1997). For the professionalization of science in England see Jim Endersby, *Imperial Nature: Joseph Hooker and the Practices of Victorian Science* (Chicago: University of Chicago Press, 2008).

47. H. Brink-Roby. "*Siren canora*: the Mermaid and the Mythical in Late Nineteenth-Century Science," *Archives of Natural History* 35:1 (2008): 1–14.

48. Ivan Sanderson. "The Wudewása or Hairy Primitives on Ancient Europe," *Genus* XVIII: 1–4(1962):109–127.

49. Ibid., 123.

50. Ivan Sanderson. "Some Preliminary Notes on Traditions of Sub-men in Arctic and Subarctic North America," *Genus* XIX:1–4 (1963):145–162.

51. Ivan Sanderson. *Abominable Snowmen: Legend Come to Life* (Philadelphia: Chilton Co., 1961): 20.

52. Ibid., 244.

53. Sanderson. *Abominable Snowmen*, 428.

54. Ivan Sanderson. "More about the Abominable Snowman," *Fantastic Universe* 2:6 (October 1959):58–64.

55. Ivan Sanderson. "The Race for Our Souls," n.d., unpublished manuscript article, APS. Sanderson did not know that the brief moment of Soviet state support of monster hunting he refers to so enthusiastically, ended quickly, and Russian monster hunters found themselves suffering and ridiculed for their work just as much as those in the West. For more on this see chapter 6.

56. Sanderson, *Abominable Snowmen*, 421.

57. Ibid., 423.

58. Willy Ley. "Do Prehistoric Monsters Still Exist?" *Mechanix Illustrated* (Feb, 1949):80–144.

59. Willy Ley. *Salamanders and Other Wonders* (New York: Viking Press, 1955): 107.

60. Willy Ley. *Willy Ley's Exotic Zoology* (New York: Viking Press, 1959): 89. This book is a combination of all three of his animal related texts.

61. P. E. Cleator to Willy Ley, August 14, 1960, box 1, Willy Ley Papers, National Air and Space Museum, Smithsonian Institution, from here NASM.

62. George Sarton to Willy Ley, May 3, 1951, box 1, folder 5, NASM.
63. Ivan Sanderson to Willy Ley, February 16, 1969, box 3, folder 4, NASM.

## 2 The Snowmen

1. Eugene S. McCartney. "Modern Analogues to Ancient Tales of Monstrous Races," *Classical Philology* 36:4 (October, 1941): 394.
2. Bernard Heuvelmans and Boris Porshnev. *L'Homme De Néanderthal est Toujours Vivant* (Librairie Plon: 1974).
3. Peter Bishop. *The Myth of Shangri-La: Tibet, Travel Writing and the Western Creation of Sacred Landscape* (Berkeley, CA: University of California Press, 1989).
4. Jerome Clark and Loren Coleman, *Cryptozoology A–Z* (New York: Fireside, 1999).
5. "Papers Relating to the Himalaya and Mount Everest," *Proceedings of the Geographical Society of London* IX (April–May 1857):345–351.
6. David L. Snellgrove. *The Cultural History of Tibet* (Orchid Press, 2006).
7. Quote in, Gardner Soule, "The World's Most Mysterious Footprints," *Popular Science* (December, 1952): 133–24.
8. "Six men-with nylon ropes-to attack Everest," *News Chronicle* (December 18, 1945). For the Everest preparations see, Churchill Archives Center, Papers of Leopold Amery, folder 14, Churchill College, Cambridge, UK.
9. Soule. "The World's Most Mysterious Footprints."
10. This information on Ripley's wartime duties taken from materials supplied by the CIA under a Freedom of Information Act request, September, 2008.
11. Ali Salim, B. Biwas, Dillon Ripley, and A. K. Gosh. *The Birds of Bhutan* (Zoological Survey of India: 2002).
12. Michel Peissel. *Tiger for Breakfast: The Story of Boris of Xathmandu* (New York: EP Dutton Co., 1966): 233.
13. This very scenario played out in the film *Abominable Snowman of the Himalayas* (1957) and the later popular children's animated Christmas special; *Rudolph the Red-Nosed Reindeer* (1964). In the film the adventurers bring nets and a cage to bring a Yeti back alive. In the cartoon the prospector, Yukon Cornelius, tries to catch the "Bumble" by tossing a net over it. In 1962 the Belgian author-illustrator Hergé (George Remi (1907–83)) sent his ubiquitous character Tin Tin to Tibet to find his friend Chang who had gone down in an airplane crash. *Tin Tin in Tibet* had the young globetrotting, journalist encounter the Yeti. On the cover Tin Tin, his partner in adventure, Captain Haddock, and a Sherpa guide are pictured encountering footprints in the snow. A stickler for authenticity, Hergé modelled the

image on the Eric Shipton photo. Hergé's Belgian compatriot Bernard Heuvelmans acted as technical consultant on several Tin Tin tales.

14. Dillon Ripley to Eric Shipton, March 13, 1953, box 45, Yeti analysis folder, Carleton Coon Papers, National Anthropological Archive, Smithsonian Institution, Washington, D.C. From here known as CCNAA.

15. John P. Jackson Jr. "In Ways Unaccademical": The Reception of Carleton S. Coon's *The Origin of the Races*," *Journal of the History of Biology* 34 (2001): 247–285.

16. Carleton Coon. *The Story of Man: from the first human to primitive culture and beyond* (New York: Alfred A. Knopf, 1954): 28.

17. He published a pair of autobiographies, *Adventures and Discoveries: The Autobiography of Carleton S. Coon* (Prentice Hall, 1981), and *A North Africa Story: The Anthropologist as OSS Agent: 1941–1943* (Gambit, 1980).

18. Continuing to suffer from his injuries, in the late 1970s Coon tried to get a disability claim based upon his wartime service. CIA records show that reviewers turned him down, saying the statute of limitations for claims had run out and because the army said he had been injured while with the OSS and technically a civilian, not part of the army, so they had no liability in the case.

19. Coon's OSS and military experiences recounted here are taken from official government personnel documents supplied by the CIA information office through a Freedom of Information Act request, November, 2008.

20. Letters between Dillon Ripley and Carleton Coon, January to March 1953, Box 45, CANAA.

21. Ralph Izzard. *The Abominable Snowman* (New York: Doubleday & Co., 1955) and Charles Stonor, *The Sherpa and the Snowman* (Hollis & Carter, 1955).

22. Coleman, *Tom Slick.*

23. Ray Miles. *King of the Wildcatters: the Life and Times of Tom Slick, 1883–1930* (College Station, TX: Texas A&M University Press, 1996).

24. Tom Slick. *Permanent Peace: A Check and Balance Plan* (Prentice Hall, 1958). The University of Texas, Austin established, with Slick estate funds, a Tom Slick Professorship of World Peace, later renamed the Tom Slick Professorship of International Affairs at their LBJ School of Public Affairs.

25. R. L. Duffus. "The Plan of Attack is on War Itself," *New York Times* (January 11, 1959): br3.

26. Greg MacGregor. "World is Asked to Accept China," *New York Times* (January 16, 1961): 3.

27. See appendix A "Tom Slick and the CIA: An Open Question," in Coleman, *Tom Slick,* 178–203. Also see; Loren Coleman, "The

Dalai Lama, Slick Denials and the CIA," in *Popular Alienation: A Steamshovel Press Anthology* (Kempton, IL: IllumiNet Press, 1995).

28. Catherine Nixon Cooke. *Tom Slick: Mystery Hunter* (Bracey, VA: Paraview Inc., 1995). Besides being Slick's niece, Cooke was also director of Slick's Mind Science Foundation.

29. Coon, *Adventures and Discoveries*, 231.

30. Ibid., 318.

31. Carleton Coon to Philip H. Wooton (*Life*), December 18, 1954, CCNAA.

32. Carleton Coon, Notes On High Altitude Project, January 13, 1955, CCNAA.

33. Ibid.

34. Jackson Jr., 2001.

35. Carleton Coon working notes, February 24, 1957, box 45, CCNAA.

36. Later Bigfoot enthusiasts made connections between flying saucers and anomalous primates.

37. Details of Coon's CIA work comes from Freedom of Information Act Records of his disability claim .

38. Coon's CIA involvement recounted here is taken from official government personnel documents supplied by the CIA information office through a Freedom of Information Act request, November, 2008.

39. Several Yeti expeditions surfaced briefly during this period. Another led by "Chicago publisher Christopher Sergell," was said to be assembling. See Mac Douglas, "The Snowman," *Everybody's Magazine* (December 6, 1958).

40. Coleman. *Tom Slick*, 194.

41. Cooke, 119.

42. "Texan Balked in Nepal Hunting," *New York Times* (October 7, 1956):10.

43. Coleman. *Tom Slick*.

44. Sara Nelson. "Yeti Evidence is 'Convincing' says Wildlife Expert Sir David Attenborough," *Daily Mail On-Line* (March 1, 2009).

45. Jamyang Wangmo. *The Lawudo Lama: Stories of Reincarnation from the Mount Everest Region* (Wisdom Publications, 2005).

46. Ibid.

47. Coleman. *Tom Slick*.

48. Vernon N. Kisling. *Zoo and Aquarium History* (James Ellis Pub., 2000).

49. William Charles Osman-Hill, "Nittaewo: An Unsolved Problem from Ceylon," *Loris: A Journal of Ceylon Wildlife* 4:1 (1945):251–262.

50. Ivan Sanderson. "More Evidence that Bigfoot Exists," *Argosy* (April 1, 1968).

51. "Soviet See Espionage in U.S. Snowman Hunt," *New York Times* (April 27, 1957):8.

52. "Soviet Scientists Trail 'Snowman,'" *New York Times* (November 16, 1958): 122.

53. "The 'Snowman' of the Pamirs," *Times*, London (January 16, 1958). Also see "Apemen May Still Be Living: Reports of Mongolian Discoveries," *Manchester Guardian* (July 12, 1958).

54. "Russians Doubt 'Snowman,'" *New York Times* (February 3, 1958): 4.

55. "Pooh-Pooh to Snowman," *New York Times* (January 10, 1960):23.

56. Valerie Vondermuhll to Carleton Coon, January 15, 1955, box 45, CCNAA. The *Saturday Evening Post* was considering underwriting the so-called "Humphrey Expedition," VonderMuhll to Coon, January 1, 1955, CCNAA.

57. George Agogino to Carleton Coon, December 30, 1958, box 8, CCNAA.

58. Ibid.

59. Photo with marginalia in box 45, CCNAA.

60. George Agogino to Carleton Coon, April 20, 1959, box 45, CCNAA.

61. Peissel, 232. Loren Coleman has his suspicions about Peissel's government connections as well. Coleman, "The Dalai Lama."

62. George Agogino to Carleton Coon, January 25, 1959, box 45, CCNAA.

63. George Agogino to Carleton Coon, May 8, 1959, box 45, CCNAA.

64. *New York Times* (December 20, 1953).

65. George Agogino to Carleton Coon," February 27, 1959, CCNAA.

66. Carleton Coon to George Agogino, March 3, 1959, CCNAA.

67. George Agogino to Carleton Coon, April 13, 1959, CCNAA.

68. Carleton Coon to George Agogino, April 17, 1959, CCNAA.

69. Adolf Schultz to George Agogino, June 6, 1959, CCNAA.

70. George Agogino to Carleton Coon, June 20, 1959, CCNAA.

71. George Agogino to Walter Krogman, January 12, 1959, CCNAA.

72. Corrado Gini. "The Scientific Basis of Fascism," *Popular Science Quarterly* 42:1 (March 1927): 99–115. Also see Giovanni Favero. *Il Fascismo Razionale: Corrado Gini fra Scienza e Politica* (Rome: Carocci, 2006).

73. Aaron Gillette. *Racial Theories in Fascist Italy* (New York: Routledge, 2002). Also see "Eugenics Conference Opens Here Today," *New York Times* (August 21, 1932): 15. When Gini was appointed a visiting lecturer at Harvard in 1935 the student council vehemently protested a fascist being welcomed at the school. "Fights Gini Appointment," *New York Times* (February 14, 1935): 4.

74. Dmitri Bayanov, "Letters in Response to 'Bigfoot' Believers," *Bigfoot Information Project* website (August 8, 2004).

75. Tom Slick. "Yeti Expedition," *Explorer's Journal* (December 1958):5–8.

76. Chapman Pincher. "Hillary Leads New Snowman Hunt," *Daily Express* (Friday, May 6, 1960). This and other newspaper clippings are

NOTES203

in the British Museum of Natural History Library, Yeti Collection, scrap book of Rosemary Powers, 1978.
77. The first episode of *Wild Kingdom* featured a segment on the Yeti.
78. "Moscow Suspicious of Hillary," *New York Times* (September 18, 1960): 45.
79. Richard Fitter. "Mask Provides New Clues to Snowman," *The Observer* (October 23, 1960).
80. Vincanne Adams. *Tigers of the Snow: and other virtual Sherpas* (Princeton: Princeton University Press, 1995): 114. Also see "Everest Headman Gives a Yeti Call in London," *Evening Standard* (December 12, 1960).
81. Ralph Izzard. *Evening Standard* (December 30, 1960) and Sir Edmund Hillary, "The Scalp is not a Scalp at all," *Sunday Times* (January 15, 1961): 5.
82. Isserman and Stewart, 352.
83. Delores Nelson, Information and Privacy Coordinator, CIA, 11/24/2008 to Brian Regal.
84. A more thorough investigation of this topic would prove most interesting.
85. George Agogino to Carleton Coon. 1/13/1961, gen. corres. A–F, 1961 file, CCNAA.

## 3  Bigfoot, the Anti-Krantz, and the Iceman

1. Don Hunter with René Dahinden. *Sasquatch/Bigfoot.* rev. ed. (Toronto: McClelland & Stewart Inc., 1993): 75.
2. Ibid.
3. Ibid.
4. For an example of the Bigfoot/UFO connections see Paul Bartholomew et al. *Monsters of the Northwoods* (New York: North Country Books, 1992).
5. For the origins of the word Sasquatch see, J. W. Burns and C. V. Tench, "The Hairy Giants of British Columbia," *Wide World Magazine* (January, 1940), and Loren Coleman, *Bigfoot! The True Story of Apes in America* (New York: Paraview Pocket Books, 2003): 31–33.
6. Andrew Genzoli, "RFD," *Humboldt Times* (October 1, 1961).
7. The description of events recounted here are taken from Dahinden, *Sasquatch* and Byrne, *The Search for Bigfoot.*
8. John Green. *On the Track of Sasquatch* (Agassiz, BC: Cheam Pub., Inc., 1968).
9. Tom Slick to Ivan Sanderson, December 11, 1959, Tom Slick folder, APS.
10. Ivan Sanderson to Jeri Walsh, May, 1961, Tom Slick folder, APS.
11. Ivan Sanderson to Albert Genzoli, 1959, Tom Slick folder, APS.
12. Lynwood Carranco, "Three Legends of Northwestern California," *Western Folklore* 22:3 (July 1963):179–185.

13. Ivan Sanderson to Jeri Walsh, May, 1961, Tom Slick folder, APS.
14. Ivan Sanderson to Tom Slick, June 24, 1961, Tom Slick folder, APS.
15. Ivan Sanderson "Memo," n.d., but probably 1962, Tom Slick folder, APS.
16. Ivan Sanderson to Tom Slick, May 28, 1962, Tom Slick folder, APS.
17. George Agogino to Carleton Coon, n.d., but from 1962, box 11, CCNAA.
18. George Agogino to Carleton Coon, November 6, 1961, box 10, CCNAA.
19. Smithsonian Torch (October 1967): 2 and M. H. Day, "In Memoriam," Journal of Anatomy 159 (1988):227–229.
20. John Napier to Roger Patterson. January 13, 1969, reproduced in Christopher Murphy, Bigfoot Film Journal (Hancock House Publishers (ebook): 2008): 75.
21. John Napier to the Trustees of the Tom Slick Foundation, August 24, 1970. John Napier papers, University College London, Special Collections, box 5, folder 22. From here referenced as UCL. Napier's wife harbored suspicions that Eric Shipton had faked the Yeti prints in his photo as a gag.
22. Lewis J. Moorman Jr. to John Napier, September 3, 1970, UCL.
23. Lewis J. Moorman Jr. to John Napier, November 11, 1970, UCL.
24. Lewis J. Moorman Jr. to John Napier, November 20, 1970, UCL.
25. John Napier to Lewis J. Moorman, December 3, 1970, UCL. My own contact with the Slick estate in April of 2008 elicited the same response.
26. John Napier. Bigfoot: the Yeti and Sasquatch in Myth and Reality (New York: E.P. Dutton & Co., 1972): 79.
27. Ibid., 162.
28. Ivan Sanderson. "The Missing Link," Argosy 368:5 (May, 1969): 23–31.
29. Quoted in Izzard, 63.
30. Don Oakley and John Lane. "Earth, Stars and Man...Ape Men and Giants," Yakima Daily Republic (November 2, 1960).
31. Brian Regal. Human Evolution: A Guide to the Debates (ABC-CLIO: Santa Barbara, 2004).
32. Michael J. O'Brien and R. Lee Lyman. Applying Evolutionary Archaeology: A Systematic Approach (New York: Springer, 2000): 117.
33. Ernest Hooton, "Pessimist's Proposal," Time (March 30, 1936).
34. Harold Sterling Gladwin. Men Out of Asia: An Exciting Picture of the Early Origins of Early American Civilization (New York: McGraw Hill, 1947): xi.
35. Ibid., 28.
36. Ibid., 30.
37. Ibid., 30.
38. Franz Weidenreich. "Giant Early Man from Java and South China," Anthropological Papers of the American Museum of Natural History

40:1 (New York: 1945). For Peking man see, Penny van Oosterzee, *The Story of Peking Man* (New York: Allen & Unwin, 2001).

39. Franz Weidenreich, *Apes, Giants and Man* (Chicago: University of Chicago, 1945), p. 41.

40. Ibid., 49.

41. Ibid., 49.

42. Weidenreich. "Giant Early Man from Java."

43. Franz Weidenreich. "Interpretations of the Fossil Material," in *Studies in Physical Anthropology: Early Man in the Far East*, W. W. Howells, ed. (American Association of Physical Anthropology, 1949):149–157.

44. Bernard Heuvelmans. "L'Homme des Cavernes a-t-il connu des Géants mesurant 3 à 4 mètres ?" *Science et Avenir* 61 & 63 (Mai, 1952).

45. Heuvelmans, *On the Track*, 98

46. Carleton Coon. *The Story of Man: from the first human to primitive culture and beyond* (New York: Alfred A. Knopf, 1954): 28.

47. Vladimir Tschernezky. "Nature of the "Abominable Snowman: A New Form of Higher Anthropoid?" *Manchester Guardian* (February 20, 1954).

48. Wladimir Tschernezky. "A Reconstruction of the foot of the Abominable Snowman," *Nature* 186:4723 (May 7, 1960): 496–97. Note that Tschernezky's name appears under at least two different spellings.

49. Heuvelmans. *On the Track*, 107.

50. Ibid., 97.

51. Ivan Sanderson. "The Missing Link," *Argosy* 368:5 (May, 1969): 23–31.

52. Party invitation to Willy Ley from Mr. and Mrs. Arthur Wang. October 15, 1968. Willy Ley Papers, NASM.

53. Party invitation from Ivan Sanderson, October 10, 1968, Ivan Sanderson Papers, Mammal Department Archive, AMNH.

54. Ivan Sanderson. "The Missing Link."

55. A copy of the Hansen Case Memo is in box 45, Yeti 1969 file, CCNAA.

56. A list of who Sanderson sent the memo to is attached to the Coon file copy.

57. Hansen Case Memo, 15.

58. Ibid.

59. Ibid.

60. Bernard Heuvelmans. "Note Preliminaire sur un Specimen Conserve dans la glace d'une forme encore inconnu d'Hominide Vivian Homo Pongoides," *Bulletin Institute Royale des Sciences Naturelles de Belgique* 45:4 (1969):1–24.

61. John Napier to Dillon Ripley. March 21, 1969, Smithsonian Institution Archives, record unit 99, box 326, Iceman folder, Dillon Ripley Papers,

Museum of Natural History, Dept V-Zoology, Division of mammals, Office of the Secretary, 1964–1971, Smithsonian Institution Archives, Washington, D.C. From here known as SMITH.

62. William Charles Osman-Hill to Dillon Ripley. February 10, 1969, SMITH.
63. Carleton Coon to Dillon Ripley. February 21, 1969, SMITH.
64. Randy Hicks to John Napier, March 27, 1969, SMITH.
65. John H. Dobkin to Wilton Dillon et al, March 13, 1969, SMITH.
66. Dillon Ripley to Frank Hansen. March 13, 1969, SMITH. Ripley sent copies of this letter to Carleton Coon, William Charles Osman Hill, Dale Stewart and Bernard Heuvelmans.
67. Frank Hansen to Dillon Ripley. March 20, 1969, SMITH.
68. John Napier to Sidney Galler. February 11, 1969, SMITH.
69. John Napier to Dillon Ripley. March 27, 1969, SMITH.
70. Dillon Ripley to J. Edgar Hoover. April 10, 1969, SMITH and " 'Ape man' escapes FBI," *Sunday Times*, London (April 27, 1969).
71. J. Edgar Hoover to Dillon Ripley. April16, 1969, SMITH.
72. Ibid., 23.
73. Ivan Sanderson, "The Missing Link," *Argosy* (May, 1969): 23–31.
74. Recollections of John Schoenherr, April, 2009.
75. Ivan Sanderson to John Napier. April 28, 1969, SMITH.
76. Marjorie Kaiman to John Napier. April 29, 1969, SMITH.
77. John Napier to Marjorie Kaiman. May 7, 1969, SMITH.
78. John Napier to Dillon Ripley. May 8, 1969, SMITH.
79. J. Lawrence Angel to E. H. Gravell. March 12, 1970. Box 126, RG155, Director, National Museum of Natural History, correspondence 1948–1970, Smithsonian Institution Archives, Washington, D.C., "The Iceman's magical mystery tour," *Sunday Times*, London, September 28, 1969, and Phil Casey, "Strange Iceman Tale: A New Race or an Old Hoax?" *Washington Post* (March 27, 1969).
80. For other miscellaneous articles about the Iceman see the British Museum of Natural History Library, Yeti Collection, scrap book of Rosemary Powers, 1978.
81. Ivan Sanderson to Ralph Izzard. June 9, 1969, Iceman file, APS.
82. Frank Hansen. "I killed the Ape-Man Creature of Whiteface," *Saga* (July 1970).
83. Napier, *Bigfoot*, 107.
84. Ivan Sanderson. "Preliminary Description of the External Morphology of What Appeared to be a Fresh Corpse of a Hitherto Unknown Form of Living Hominid," *Genus* XXV: 1–4 (1969): 249–84.
85. K. Stolyhwo. "Le crâne de Nowosiolka considéré com preuve de l'existence à l'époque historique de formes apparentées à *H. primigenius*," *Bulletin International de l'Académie des Sciences de Cracovie* (1908):103–26.
86. Edward Tyson. *Anatomy of a Pygmie* (London: Thomas Bennet, 1699). The full title is *Orang-Outang, sive Homo Sylvestris. Or,*

*the anatomy of a Pygmie compared with that of a Monkey, an Ape, and a Man. To which is added, A Philosophical Essay concerning the Pygmies, the Cynocephali, the Satyrs, and the Sphynges of the ancients.*

87. Bernard Heuvelmans and Boris Porshnev. *L'Homme de Néanderthal es ToujoursVivant* (Paris: Plon, 1974). This is one of the few of Heuvelmans' book not translated into English.
88. John Napier to Dillon Ripley. June 2, 1969, SMITH.
89. John Napier to Dillon Ripley. February 16, 1971, box 519, SMITH.

## 4  The Life of Grover Krantz

1. "Christianity and Krantz," dated 1952, Scrapbook of Ester Marie Krantz, box 12, oversize, NAA. Years later Krantz grappled with creationist Duane Gish who, like Krantz, was a Berkeley alumnus.
2. Matthew Bowman. "A Mormon Bigfoot: David Patten's Cain and the Conception of Evil in LDS Folklore," *Journal of Mormon History* 33:3 (Fall 2007): 62–82 and Shane Lester. *Clan of Cain: the Genesis of Bigfoot* (self published, 2001).
3. GSK diary entry, 1/1/1949. folder 1577, box 15. NAA.
4. Krantz preserved the ragged photo to the end of his life and included it in his papers.
5. GSK resume in folder 0001, box 1, NAA.
6. Ibid.
7. Grade information taken from official GSK transcripts, registrar, University of California, Berkeley.
8. GSK resume, folder 001, box 1, NAA.
9. GSK, "Sphenoidal Angle and Brain Size," *American Anthropologist* 64:3 (1962):521–23.
10. GSK. *Only a Dog*, (Wheat Ridge, CO: Hoflin Pub., 1998).
11. Ibid., 7.
12. Ibid., 9.
13. Ibid.
14. GSK to Roger Patterson, 8/22/1970, folder 0343, box 7, NAA.
15. See a notebook of drawings folder 1559, box 14, NAA.
16. GSK resume, folder 001, box 1, NAA.
17. *Daily Humboldt Times*, 10/15/1958.
18. Wolfgang Saxon. "Sherwood Washburn, Pioneer in Primate Studies Dies at 88," *New York Times* (April 9, 2000). Also see, Sherwood Washburn and Irven DeVore, "Social Behavior of Baboons and Early Man," in Sherwood Washburn ed. *Social Life of Early Man* (Aldine Pub. Co., 1961): 91–105.
19. Letter of reference, and other materials, from E. Adamson Hoebel, January 19, 1968. Washington State University, Office of Procedures, Records, and Forms. From here known as WSU.
20. Ibid.

21. Ivan T. Sanderson. "First Photos of "Bigfoot," California's Legendary "Abominable Snowman,"" *Argosy* 336:2, February, 1968:23–29, 72, 127–128.
22. "U Lecturer from West Has Hunted 'Snowman,'" *Minneapolis Star*, 1/25/1968.
23. Ibid.
24. GSK to Robert Littlewood, November 17, 1967, WSU.
25. Reference letter from Robert F. Spencer, January 15, 1968, WSU.
26. See folder 0403, box 8 and Krantz curriculum vita, folder 0001, box 1, NAA.
27. John Green. *On the Track of Sasquatch* (Agassiz, BC: Cheam Pub., Inc., 1968): 74.
28. Brian Regal. "Entering Dubious Realms: Grover Krantz, Science, and Sasquatch." *Annals of Science* 66:1 (January 2009): 83–102. 90.
29. Green. *On the Track of Sasquatch*, p. 74.
30. *Denver News*, 3/5/1988, p.1A and 10A.
31. "Cold freezes hounds off humanoid's trail," *Montana Standard* (12/7/1969).
32. Daegling, 81.
33. Krantz recollected that it was John Green himself who had covered a number of the better Bossburg prints, but was not sure. Krantz was also a bit hazy on just when he arrived at Bossburg, though internal evidence seems to suggest mid-December rather than January of 1970. Richard Noll, "Interview with Dr. Grover Krantz," *Bigfoot Encounters.com* (July 1, 2001).
34. GSK, "Sasquatch Handprints," *North American Research Notes* 5:2, Fall, 1971:145–151.
35. GSK, *Big Footprints*.
36. Porshnev, Boris. "Troglodytidy i gominidy v sistematike i evolutsii vysshikh primatov," *Doklady Akademii Nauk SSSR*, 188:1 (1969). This article was later published in an English translation as "The Troglodytidae and the Hominidae in the Taxonomy and evolution of higher primates," *Current Anthropology* 15:449, 1974:450.
37. John Green. *Year of the Sasquatch* (Agassiz, BC: Cheam Pub. Ltd., 1970): 35.
38. Roderick Sprague. "Editorial," *Northwest Anthropological Research Notes* 4:2 (Fall, 1970): 127–128. In 2002 the journal's title changed to *Journal of Northwest Anthropology [JONA]*.
39. Grover Krantz. "Sasquatch Handprints," *North American Research Notes* 5:2 (Fall, 1971):145–51.
40. W. Tschernezky. "A Reconstruction of the foot of the 'Abominable Snowman," *Nature* 186:4273 (May 7, 1960): 496–97. Tschernezky had already put forward the *Gigantopithecus* theory in an article for the *Manchester Guardian* in 1954.
41. Krantz. *Bigfoot Sasquatch*, 54.
42. GSK. "Anatomy of a Sasquatch Foot," *North American Research Notes* 6:1 (Spring, 1972): 91–104.

43. Aaron Gillette. *Eugenics and the Nature-Nurture Debate in the Twentieth-Century* (New York: Palgrave-Macmillan, 2007).
44. GSK. "Anatomy of a Sasquatch Foot."
45. For eugenics and typology see, George W. Stocking, Jr., ed. *Bones, Bodies, Behavior: Essays on Biological Anthropology* (University of Wisconsin Press: Madison, WI, 1988).
46. Milford Wolpoff and Rachel Caspari. *Race and Human Evolution: A Fatal Attraction* (New York: Simon & Schuster, 1997).
47. Carleton Coon. *The Races of Europe* (Macmillan, New York: 1939) and *Origin of the Races* (New York: Knopf, 1962).
48. Carleton Putnam. *Race and Reason: a Yankee View* (Washington, D.C.: Public Affairs Press, 1961). Putnam is still held in reverence as a 'scholarly' author by reactionary right wing pundits and racialists as is Coon.
49. Aaron Gillette. *Eugenics and the Nature-Nurture Debate in the Twentieth-Century* (New York: Palgrave-Macmillan: 2007): 147, 162.
50. Pat Shipman. *The Evolution of Racism: Human Differences and the Use and Abuse of Science* (Cambridge, MA: Harvard University Press, 1994): 173.
51. Sherwood Washburn. "The Study of Race," *American Anthropologist* 65 (1963):521–31.
52. Theodosius Dobzhansky. "Genetic Entities in Hominid Evolution," in Sherwood Washburn ed. *Classification and Human Evolution* (Chicago: Aldine Pub. Co., 1963): 361.
53. Grover Krantz. "Review of *Human Variation: Races, Types, and Ethnic Groups*," by Stephen Molnar, *American Anthropologist* 85 (1983):702.
54. Carleton Coon. "Why There has to be a Sasquatch," in Markotic, Vladimir and Grover Krantz eds., *The Sasquatch and Other Unknown Hominoids* (Calgary: Western Publishers, 1984).
55. Krantz was so taken by these discussions with Coon that he hurriedly wrote up notes afterward so he could keep a record of them. See folder 0428, box 9, NAA.
56. Coon to GSK, 3/19/1977, folder 0336, box 7, NAA. Coon's work appears in the bibliographies of several of Krantz's books and papers.
57. See GSK, folder 0403, box 8, NAA.
58. See GSK, folder 0433, box 10, NAA.
59. See GSK, folder 0316, box 6, NAA.
60. Will Duncan. "What is Living in the Woods, and Why it Isn't Gigantopithecus," in Craig Heinselman ed. *Crypto Hominology Special #1* on-line (April 7, 2001).
61. See Everett Ortner, "Do 'Extinct' animals still survive?" *Popular Science Monthly* (1959), Don Oakley and John Lane, "Earth, Stars and Man: Ape men and Giants," *Yakima Daily Republic* (November 2, 1960) and Willy Ley, *Exotic Zoology* (New York: Viking Press, 1959).

62. Don Oakley and John Lane, "Earth, Stars and Man: Ape men and Giants," *Yakima Daily Republic* (November 2, 1960).

63. Michael Grumley. *There Were Giants in the Earth* (Garden City, NY: Doubleday & Co., Inc., 1974): 91.

64. Ibid., 90.

65. B. Ann Slate and Alan Berry. *Bigfoot* (New York: Bantam, 1976): xiii.

66. Ibid., 37.

67. Manuscript titled "History," folder 0344, box 7, NAA.

68. Bernard Heuvelmans to GSK. September 13, 1973, folder 0340, box 7, NAA.

69. GSK. *The Process of Human Evolution* (Cambridge, MA: Schenkman, 1981).

70. GSK. "*Homo erectus* Brain Size by Subspecies," *Human Evolution* 10:2, 1995:107–117.

71. GSK. *The Origins of Man* (University of Minnesota, UMI Dissertation Services, 1971), 13.

72. GSK. "Pithecanthropine Brain Size and its Cultural Consequences," *Man* 61, May, 1961:85–87, "Brain Size and Hunting Ability in Earliest Man," *Current Anthropology* 9:5, December, 1968:450–451, and "Sapienization and Speech," *Current Anthropology.* 21:6, December, 1980:773–792.

73. GSK. *Climatic Races and Descent Groups* (North Quincy, MA.: Christopher Publishing House, 1980), 231.

74. Wolpoff's first scholarly article on regional continuity was Thorne, A. G., and M. H. Wolpoff, "Regional continuity in Australasian Pleistocene hominid evolution." *American Journal of Physical Anthropology* 55 (1981):337–349.

75. See Richard Noll, "Interview with Dr. Grover Krantz," *Bigfoot Encounters.com* (July 1, 2000).

76. The only reference Krantz ever made to Wolpoff was a single reference to an article for his introduction to *The Scientist Looks at the Sasquatch*, Roderick Sprague and GSK eds. (Moscow, Idaho: University Press of Idaho, 1977): 26.

77. For his views on how the aboriginal people entered the Americas—and presumably Sasquatch as well—see GSK, "The Populating of Western North America," *Method and Theory in California Archaeology* 1, 1977:1–63.

78. Interview with Milford Wolpoff, July 13, 2010.

79. Interview with Milford Wolpoff, December 10, 2008.

80. GSK classroom notes for Anthr 465/565 Human evolution, WSU. (1991–2000). Folder 0439, box 10, NAA.

81. GSK. *Process of Human Evolution*, 173.

82. GSK classroom notes for Anthr 465/565 Human evolution, WSU. (1991–2000). Folder 0439, box 10, NAA.

83. GSK. *Process of Human Evolution*, 461.

84. Ibid., 466.
85. Ibid., 443.
86. Quoted in Robert Sullivan, "Bigfoot," *Open Spaces Quarterly* 1:3 (1999).
87. A. Adamson Hoebel. *Man in the Primitive World* (McGraw-Hill: New York, 1949):30.
88. GSK's copy of Hoebel's *Man in the Primitive World* with marginalia from collection of B. Regal, 30.
89. GSK. *Bigfoot-Prints*, 12.
90. For Krantz's marriages see his will and testament for 1982, folder 0406, box 8, NAA.
91. In his final will, Krantz left his own skeleton, along with Clyde, to the Smithsonian Institution.
92. GSK. *Only a Dog*, (Wheat Ridge, CO: Hoflin Pub., 1998). Krantz left Clyde's and his own body to the Smithsonian Institution's osteology collection. In 2009 the Smithsonian mounted their skeletons in a pose mimicking a photo of the two. Krantz and Clyde would be able to stay together for eternity.
93. Ibid., 32.

## 5 Suits and Ladders

1. Ivan Sanderson. "First Photos of Bigfoot: California's "Abominable Snowman,"" *Argosy* 29 (February, 1968).
2. Murphy, Christopher. *Bigfoot Film Journal* (Blaine, WA: Hancock House Publishers (ebook): 2008).
3. This description of events is based upon Sanderson, "First Photos," and American Museum of Natural History mammal department archive records.
4. Greg Long. *The Making of Bigfoot: the inside story* (Amherst, NY: Prometheus Books, 2004).
5. Ivan Sanderson. "The Strange Story of America's Abominable Snowman," *True* 40:271 (December 1959):40–126.
6. Roger Patterson. *Do Abominable Snowmen of America Really Exist?* (Yakima, WA: Franklin Press, 1966): viii.
7. John Green. *Sasquatch: the Apes Among Us* (Blaine, WA: Hancock House Publishers: 2006):114–115.
8. Hunter and Dahinden. *Sasquatch/Bigfoot*, 112.
9. This sequence of events comes from Daniel Perez, *Bigfoot at Bluff Creek* (Center for Bigfoot Studies: Norwalk, CA, 2003). Perez interviewed all the parties involved to get probably the most accurate overall description of the events: for those who believe the film genuine.
10. Ibid.
11. Hunter and Dahinden. *Sasquatch/Bigfoot*.
12. Ibid., 116.

13. Ivan Sanderson. "The Patterson Affair," *Pursuit* (June, 1968). *Pursuit* was the newsletter of Sanderson's Society for the Investigation of the Unexplained.

14. Ivan Sanderson, undated typescript article, "Man-Things," Ivan Sanderson Papers, American Philosophical Society. From here APS. Note that the opening quote to this chapter also comes from this mss.

15. Interview with Joe Davis. April, 2009.

16. Meldrum. *Sasquatch.*

17. Ivan Sanderson. "First Photos of Bigfoot: California's 'Abominable Snowman,'" *Argosy* 29 (February, 1968).

18. Alex Faulkner, "US Film of Abominable Snowman," *The Daily Telegraph,* London (November 22, 1967).

19. For the publishing information see, New York Public Library, Rare Book and Manuscript Room, Popular Publications Collection, box 50, index card file.

20. Charles Fort. *Book of the Damned* (1919).

21. Interview with Joe Davis, April, 2009.

22. Joshua Blu Buhs. *Bigfoot: Life and Times of a Legend* (Chicago: University Of Chicago Press, 2009):110.

23. See note 14.

24. Ivan Sanderson to Hobart Van Deusen, January 7, 1962, Ivan Sanderson file, Mammalogy Department Archive, AMNH.

25. Materials in the Ivan Sanderson file, Mammalogy Department Archive, AMNH.

26. For a partial list of Sanderson's contributions to *Argosy* see the Popular Publication Collection, NYPL.

27. Dick Kirkpatric. "The Search for Bigfoot: Has a 150 Year-Old Legend Come to Life on this Film?" *National Wildlife* (April–May 1968):43–47.

28. *Radio Times London,* television listings (July 25, 1968).

29. Van Deusen's SITU membership card (#383H) is in the Ivan Sanderson file, Mammalogy Department Archive, AMNH.

30. Sydney Anderson to Peter Byrne, August 24, 1979, "Bigfoot Project" folder, Department of Mammalogy Archives, AMNH.

31. David C. Anderson. "Stalking the Sasquatch," *New York Times* (January 20, 1974): 231.

32. Sydney Anderson to James Spink, February 11, 1976, AMNH.

33. Rex Nelms to Sydney Anderson, April 30, 1976, AMNH.

34. Susan Hassler to Sydney Anderson with his reply mss marginalia, March 9, 1977 (reply sent March 15), AMNH.

35. "Bigfoot Photo Baffles Experts!" *Weekly World News* (April 29, 1980).

36. Ibid.

37. Paul Bartholomew to Sydney Anderson, 4/24/1980, AMNH.

38. Sydney Anderson to Paul Bartholomew. 5/26/1980, AMNH.

39. See correspondence in the Napier Papers, box 5, folder 22, UCL.

40. Manuscript reply, John Napier to BBC Wildlife, Napier Papers, box 5, folder 22, UCL.
41. Green. *On the Track of Sasquatch.*
42. GSK. "Additional Notes on Sasquatch Foot Anatomy," *North American Research Notes* 6:2 (Fall, 1972):230.
43. Ibid., 236. Sheldon worked with E.A. Hooton and Carleton Coon at Harvard. Patricia Vertinsky, "Physique as Destiny: William H. Sheldon, Barbara Honeyman Heath and the Struggle for Hegemony in the Science of Somatotyping," *CBMH/BCHM* 24:2 (2007):291–316.
44. GSK notebooks from June, 1981, folder 0326, Box 6, NAA.
45. Grover Krantz. *Big Footprints: a Scientific Inquiry into the Reality of Sasquatch* (Boulder, CO: Johnson Books, 1992).
46. GSK notebooks from June, 1981, folder 0326, Box 6, NAA.
47. Kodac K100 film camera instruction manual: 4.
48. Peter Byrne to GSK. 5/18/1993 and 5/19/1993, folder 0325, Box 6, NAA. Yakima Police report #67–8923.
49. Peter Byrne to GSK. 5/18/1993, folder 0325, Box 6, NAA.
50. Rusty Dornin. "Don't Believe in Aliens? Visit San Francisco's UFO Museum," *CNN Interactive* (April 19, 1997).
51. Loren Coleman and Patrick Huyghe, "Letter to the Editor of *Skeptical Inquirer*" (January 26, 2000). Listed on the webpage, "Letters *Skeptical Inquirer* Refused to Publish."
52. GSK. *Big Footprints*, 119.
53. Green. *Sasquatch*, 119.
54. David Wasson. *Yakima Herald-Republic* (January 31, 1999). Also see, Superior Court of Washington for Yakima County, February 6, 1976, Gimlin vs. DeAtley and Patterson (short title) no. 58594.
55. Green. *Sasquatch*, 123.
56. AAG Jennifer Hubbard Geller, Washington Attorney General's Office Memorandum, February 26, 1996, folder 0343, box 7, NAA
57. Greg Long. *The Making of Bigfoot: the inside story* (Amherst, NY: Prometheus Books, 2004).
58. Kal K. Korff, "The Making of Bigfoot," *Fortean Times* 119 (February, 2005):34–39. The quote is from page 34.
59. *ESkeptic*, "Bigfoot Big Con," and Michael Dennett and Daniel Loxton, "Some reasons for Caution about the Bigfoot Film Expose," *Skeptical Inquirer* (Jan–Feb 2005).
60. Korff. 39.
61. Kay Bartlett. "Bigfoot Hunters Don't Get Along," *Times-Union and Journal*, Jacksonville, Florida (January 28, 1979): A1–A4.
62. Peter Byrne. "Robert 'Bob' Titmus: Bigfoot Expert, Veteran Woodsman, Master Tracker," (Bigfoot Encounters website, 2009).
63. Phil Busse, "Looking for Mr. Bigfoot: The Western Bigfoot Society and the Eternal Search for Truth," *Portland Mercury* (September 14, 2000).

64. Peter Byrne. *The Search for Bigfoot: Monster, Myth or Man?* (Washington, DC: Acropolis Books, 1975): 138.
65. Ibid., 142.
66. Sydney Anderson to Peter Byrne, August 24, 1979, Peter Byrne folder, AMNH.
67. Bob Downing. "Chief Hunter of Bigfoot Calls it Quits," *San Francisco Examiner and Chronicle* (September 30, 1979): B:5.
68. Long. 109–10.
69. Peter Byrne to Sydney Anderson, October 16, 1979, Peter Byrne folder, AMNH.
70. Long. 187.
71. Bernard Heuvelmans, notes, folder 0340, box 7, NAA.
72. Bernard Heuvelmans to GSK, December 25, 1991, folder 0340, box 7, NAA.
73. Quoted in Long, 193.
74. Ivan Sanderson to Alma Sanderson, UD, Sanderson, Alma folder, APS.
75. Ivan T. Sanderson to Ivan L. Sanderson, February 1, 1968, Sanderson family folder, APS.
76. Ivan Sanderson to Bursar, Trinity College, Cambridge University, November 28, 1968, University of Cambridge folder, APS.
77. Ivan Sanderson to Ralph Izzard, April 21, 1972, Izzard, Ralph folder, APS.
78. Mark A. Hall. "Biography of Ivan Sanderson," *Wonders* (December 1992): 65–67.
79. Ivan T. Sanderson to Ivan L. Sanderson, February 1, 1968, Sanderson family folder, APS.
80. William Montagna, "From the Director's Desk," *Primate News* (September 1976):7–9.
81. Ibid.
82. Green. *Sasquatch*, 129.
83. "Bob Gimlin to be Special Guest at 2009 Texas Bigfoot Conference," *Cryptomundo* web site.

## 6  The Problems of Evidence

1. Dmitri Bayanov. *America's Bigfoot: Fact, Not Fiction* (Moscow: Crypto Logos, 1997):55.
2. GSK. "*Homo erectus* Brain Size by Subspecies," *Human Evolution* 10:2, 1995:107–17, "Pithecanthropine Brain Size and its Cultural Consequences," *Man* 61, May, 1961:85–87 and *Big Footprints: A Scientific Inquiry Into the Reality of Sasquatch* (Boulder, CO: Johnson Books, 1992).
3. Ibid.
4. M. Estellie Smith. "Review of *Climatic Races*," *Annals of the American Academy of Political and Social Science* 453 (January, 1981): 290–91.

5. Robert B. Eckhardt. "Review of *Climatic Races*," *Amer. Anth.* 84:2 (June, 1982):454–56.

6. David Frayer, Milford H. Wolpoff, Alan G. Thorne, Fred H. Smith, and Geoffrey G. Pope. "Resolving the Archaic-to-Modern Transition: A Reply," *Amer. Anth.* 96:1, March, 1994:152–55.

7. GSK class notes in folder 0419, box 9, NAA.

8. GSK. *Process of Human Evolution*, p. 238.

9. E. L. Simons and S. R. K Chopra. "Gigantopithecus (Pongidae Hominiodea) A New Species from North India," *Postilla* 138 Peabody Museum, Yale University (October 1, 1969): 1.

10. Wen Chung Pei. "Giant Ape's Jawbone Discovered in China' " *Amer. Anth.* 59:5 (October, 1957): 834–38, and "Excavation of Liucheng *Gigantopithecus* cave and exploration of other caves in Kwangsi," Institute of Vertebrate Paleontology and Paleoanthropology, *Academica Sinica* 7, (Peking: Science Press, 1965).

11. Ibid., p. 836.

12. David Pilbeam. "Gigantopithecus and the Origins of Homimidae," *Nature* 225 (February 7, 1970): 516–19.

13. Bruce R. Gelvin. "Morphometic Affinities of Gigantopithecus," *American Journal of Physical Anthropology* 53:4 (1986): 541–568.

14. Napier. *Bigfoot*, 117.

15. Lonnie Somer. "New Signs of Sasquatch Activity in the Blue Mountains of Washington State," *Cryptozoology* 6 (1987):65–70; and Michael R. Dennett. "Bigfoot evidence: are these tracks real?" *Skeptical Inquirer* (September 22, 1994).

16. GSK notes in folder 0333, box 7, NAA.

17. Krantz included this line of thinking in *Big Footprints: A Scientific Inquiry Into the Reality of Sasquatch* (Boulder, CO: Johnson Books, 1992).

18. Boris Porshnev. "The Troglodytidae and the Hominidae," *Curr. Anth.* 15:449, 1974:450.

19. Bernard Heuvelmans to GSK, 8/13/1985, folder 0340, box 7, NAA.

20. Sir Peter Scott and Robert Rines. "Naming the Loch Ness Monster," *Nature* 258 (December 11, 1975):466–468.

21. Edward Bousefield and Paul LeBlond. "An account of *Cadborosaurus willsi*, new genus, new species, a large aquatic reptile from the pacific coast of North America," *Amphipacifica Journal of Systematic Biology* *1 supp.* 1 (1995): 1–25.

22. Van Valen and Krantz were friends and both members of the International Society of Cryptozoology.

23. Leigh Van Valen to GSK, 1/21/1986, folder 0333, box 7, NAA. See same folder for the reviewer's notes. The paper was eventually published as "A Species Named from Footprints" in the Krantz-friendly *Northwest Anthropological Research Notes* 19:1 (Spring, 1985): 93–99.

24. Russell Ciochon, John Olsen and Jamie James. *Other Origins: the search for the great apes in human prehistory* (New York: Bantam Books, 1990), p. 228.

25. Quoted in Dennett. "Bigfoot Evidence," p. 500.

26. GSK. *Bigfoot Sasquatch Evidence*, p. 63.

27. Ibid.

28. Michael Dennet. "Bigfoot Jokester Reveals Punchline—Finally," *The Skeptical Inquirer* 7:1 (Fall, 1982): 8–9, "Evidence for Bigfoot? An Investigation of the Mill Creek 'Sasquatch Prints,'" *The Skeptical Inquirer* 13:3 (Spring 1989): 264–272, and "Bigfoot Evidence: are these tracks real?" *The Skeptical Inquirer* 18:5 (Fall, 1994): 498–508. Also see, David Daegling, *Bigfoort Exposed: An Anthropologist Examines America's Enduring Legend* (Walnut Creek, CA: AltaMira Press, 2004).

29. René Dahinden to GSK. 1/3/1986, folder 0342, box 7, NAA.

30. John Robinson to Ed Palma. 12/22/1982, folder 0333, box 7, NAA.

31. See correspondence between Ripu Singh and GSK, folder 0330, box 6, NAA.

32. A. G. de Wilde to GSK. 1/3/1984, folder 0332, box 6, NAA.

33. John Berry to GSK. 9/3/1984, folder 0331, box 6, NAA.

34. See folder 0317, box 6, NAA.

35. Richard Greenwell to Alex Roche. 9/13/1985, folder 0333, box 7, NAA.

36. John Berry and Stephen Haylock. "The Sasquatch Foot Casts," *Fingerprint Whorld* (January, 1985):59–63.

37. The list of such works is quite large. Besides Heuvelmans and Sanderson, see Green, "What is a Sasquatch?" in *Manlike Monsters*.

38. For descriptions of fossil hominid evidence see, Ian Tattersall, Eric Delson, and John Van Couvering. *Encyclopedia of Human Evolution and Prehistory* (Garland Publishing: New York, 1988).

39. "Soviet Scientists Trail Snowman," *New York Times* (November 16, 1958):122.

40. "Pooh-pooh to Snowmen," *New York Times* (January 10, 1960): 23.

41. "Russians Doubt Snowman," *New York Times* (February 3, 1958): 4.

42. Boris Porshnev and A. A. Shmakov, eds. *Informatsionnye Materialy Komissii po Izucheniyu Voprosa o "Snezhnym Chelo-veke,"* I–IV, Moskva (1958–59).

43. Porshnev's name is also found spelled Porchnev.

44. Boris Porshnev. "The Problem of Relic Paleoanthropus," *Soviet Ethnography* 2 (1969): 115–30.

45. "Engels on the Origin and Evolution of the Family," *Population and Development* Review 14:4 (December, 1988):705–729, quote on 706.

46. Professor L. Astanin to GSK. July 1, 1972, folder 0347, box 7, NAA.

47. Myra Shackley. "Case for Neanderthal Survival," *Antiquity* LVI (1982):31–41, and P. R. Rinchen. "Almas Still Exist in Mongolia," *Genus* 20 (1964):188–92.

48. The translation of Mongolian and Russian family names into English can present confusion as there are often several versions found in

the literature. I will follow the spelling as it appears in an article by Rinčen in French.

49. Michael Heaney. "The Mongolian Almas: a Historical Reevaluation of the Sighting by Baradiin," *Cryptozoology* 2 (1983): 40–52.
50. Myra Shackely. *Still Living? Yeti, Sasquatch and the Neanderthal Enigma* (New York: Thames and Hudson, 1983).
51. Heaney. "The Mongolian Almas."
52. Ibid.
53. Yöngsiyebü Rinčen. "L'Heritage Scientifique du Prof. Dr. Zamcarno," *Central Asiatic Journal* 4:3 (1959):199–206.
54. "Important Find of Skulls," *Manchester Guardian* (July 12, 1958).
55. V. Rinčen. "Almas Still Exist in Mongolia," *Genus* 20 (1964): 186–192.
56. Shackley. *Still Living*, 99.
57. Rinčen. "L'Heritage Scientifique."
58. Heany, 45.
59. "Soviet scientist reports he may have seen 'Abominable Snowman,'" *Sarasota Herald-Tribune* (January 19, 1958):3.
60. See Ian Tattersall, Eric Delson, and John Van Couvering. *Encyclopedia of Human Evolution and Prehistory* (New York: Garland Publishing, 1988) and Christopher Stringer and Robin McKie. *African Exodus: the Origins of Modern Humanity* (New York: Henry Holt, 1996).
61. The discovery of the controversial *Homo floresiensis*, on the island of Flores in Indonesia in 2003, seemed to confirm a surviving hominid group possible. Even if this theory is correct *Homo floresiensis* went extinct about thirteen thousand years ago, well before modern times. See Gregory Forth, "Hominids, hairy hominoids and the science of humanity," *Anthropology Today* 21:3 (June 2005):13–18.
62. Boris Porshnev. "The Troglodytidae and the Hominidae in the Taxonomy and evolution of higher primates," *Current Anthropology* 15:449, (1974):450.
63. *The Times*, London (January 16, 1958).
64. Bernard Heuvelmans and Boris Porchnev. *L'Homme De Néanderthal est Toujours Vivant* (Librairie Plon 1974).
65. Heuvelmans and Porshnev, 171–77.
66. Shackley, 113.
67. Heuvelmans and Porshnev, 164–65.
68. Emanuel Vlček. "Old Literary Evidence for the Existence of the 'Snowman' in Tibet and Mongolia," *Man* 59 (August, 1959): 133–34.
69. Ibid., 134.
70. In Vlček's article, he calls him B. Rinchen.
71. Ibid.
72. Reuters News Service (March 11, 1992) and Bryan Stevenson. "On the Trail of Sasquatch," *True Fortune* (December, 1975).
73. "Hunting the Almasty," *Economist* (June 26, 1992) and "French, Russians to Hunt Caucasian Yeti," Reuters News Service (March 11, 1992).

74. Linda Coil Suchy. *Who's Watching You? An Explanation of the Bigfoot Phenomenon in the Pacific Northwest* (:Blaine, WA: Hancock House Publishers, 2009): 283.
75. Dmitri Bayanov. *America's Bigfoot: Fact, not Fiction*, (Moscow: Crypto Logos, 1997): 62.
76. Dmitri Bayanov. *Bigfoot Research: The Russian Vision* (Moscow: Crypto-Logos, 2007).
77. Lloyd Pye. "Response to the 'Russian Bigfoot' Episode on National Geographic's Cable Television Show 'Is It Real?'" (LloydPye.com 2007).
78. Dmitri Bayanov. *America's Bigfoot: Fact, Not Fiction* (Moscow: Crypto Logos, 1997).
79. Bayanov. *Bigfoot Research: The Russian Vision*, XIII.
80. Green and Coy self-published *50 Years with Bigfoot*. For details see Kristin Luna. *Tennessee Curiosities: Quirky Characters, Roadside Oddities & Other Offbeat Stuff* (Connecticut: Globe Pequot, 2010): 79. There is a body of literature on long term personal encounters with anomalous primates. See Jan Klement. *The Creature: Personal Experiences with Bigfoot* (Elgin, PA: Allegheny Press, 2006), and Sali Sheppard-Wolford, *Valley of the Skookum: Four Years of Encounters with Bigfoot* (Enumclaw, WA: : Pine Winds Press, 2006) for just two of the better known versions. For an overall assessment of the genre see, Thom Powell. *The Locals: A Contemporary Investigation of the Bigfoot/Sasquatch Phenomenon* (Blaine, WA: Hancock House, 2003).
81. Jill Thomas. "Russian Researcher Visits Overton County Tennessee," *Herald Citizen* (September 24, 2004).
82. Dmitri Bayanov. *Is Manimal More than Animal?* (International Center for Hominology, Moscow: 2006).
83. Dmitri Bayanov to GSK. March 8, 1985, folder 0346, box 7, NAA.
84. Bayanov, *America's Bigfoot*, 23.
85. Ibid., 31.
86. Richard Greenwell to GSK. 5/27/1991, folder 0235, box 6, NAA.
87. Dmitri Bayanov to Richard Greenwell. 5/2/1991, folder 0326, box 6, NAA.
88. *American Anthropologist* to GSK. 6/23/1979, folder 0243, box 4, NAA.
89. John Fleagle to GSK. 7/23/1997, folder 0253, box 4, NAA.

## 7   A Life with Monsters

1. GSK. *Bigfoot/Sasquatch*, p. 86.
2. Quoted in Michael Schmeltzer, "Bigfoot Lives," *Washington Magazine* V (Sept–Oct, 1998): 64–69.
3. Quoted in Bill Loftus, "Professor's Sasquatch Hunt a Private Matter," *Lewiston Morning Tribune* (April 4, 1988).

4. Carleton Coon. Review of "Sapienization and Speech," Current Anthropology (April 23, 1980).

5. Bill Hill. "Seeing the Sasquatch and the Tooth Fairy," *Lewiston Morning Tribune* (March 6, 1988).

6. See NAA for letters from students supporting GSK's tenure and promotion. For the quote, see Stringer and McKie, *African Exodus*, 91.

7. *Seattle Times* (February 13, 1972).

8. Quotes from an interview with Robert Ackerman, January 15, 2008.

9. GSK to Bernard Heuvelmans. October 12, 1982, folder 0340, box 7, NAA.

10. Bernard Heuvelmans to GSK. August 9, 1972, folder 0340, box 7, NAA.

11. *Denver News* (May 5, 1988): 1A & 10A.

12. Quoted in Schmeltzer, 64.

13. Undated AP article, likely from late 1980s, in folder 0323, box 6, NAA.

14. File box 9, folder 0409, NAA.

15. Official WSU records, Office of the Registrar.

16. GSK to Allan H. Smith. March 4, 1977, WSU.

17. Allan H. Smith to GSK. March 8, 1977, WSU.

18. Interview with John Bodley, February, 2010, and GSK faculty records, Official WSU records, Office of the Registrar.

19. *East Washingtonian*, Pomeroy, WA. 8/16/1971.

20. Richard Noll. "Interview with Dr. Grover Krantz," *Bigfoot Encounters. com* (July 1, 2001).

21. "Hunting Bigfoot by Air," *AP News Service* article with marginalia, 24 February 1988, folder 0323, box 6, NAA.

22. "Professor: Track Down, Kill a Bigfoot," *Salt Lake Tribune* January 22, 1996.

23. Form letter in folder 0323, box 6, NAA.

24. GSK, box 6, NAA.

25. Dwight G. Smith and Gary Mangiacopra, "What Readers Wrote In: secondary Bigfoot sources as given in the Letters-to-the-Editors column of the 1960s–1970s Men's Adventure Magazines," *North American BioFortean Review* 5:4 issue 13 (December, 2003): 19–31.

26. GSK, box 6, NAA.

27. Kirsten Francis to GSK. Undated but from the 1970s, box 6, folder 0321, NAA.

28. Kirsten Francis, Cheri, Bernard, Anne and Stefanie Johnson to GSK. Undated but from the 1970s, box 6, folder 0321, NAA.

29. Anne Marie Costa to GSK. Undated but from the 1970s, box 6, folder 0321, NAA.

30. Paul Houghton to GSK. August 2001, box 3, folder 0216, NAA.

31. For the Jacko story, see, Jerome Clark and Loren Coleman. *Cryptozoology A–Z* (New York: Fireside, 1999).

32. Chilco Choate to GSK. March 26, 1970, box 6, folder 0344, NAA.
33. Robert A. Stebbins, "The Amateur," *Pacific Sociological Review* 20:4 (October 1977): 588.
34. For criticism on Meldrum's work see; Jesse Harlan Alderman, "Idaho Prof Criticized Over Bigfoot Study," *AP* wire story (November 3, 2006).
35. In 1974 Krantz offered an unofficial seminar on Sasquatch at WSU. Faculty complaints insured it never ran again.
36. For Stebbins's work on leisure see; Robert A. Stebbins. "Serious Leisure: A Conceptual Statement," *Pacific Sociological Review* 25:2 (April, 1982): 251–72, "The Amateur," *Pacific Sociological Review* 20:4 (October 1977): 588, *Amateurs: Margin Between Work and Leisure* (Beverley Hills, CA: Sage Publications, 1979), and *Serious Leisure: a Perspective for our Time* (New Brunswick, NJ: Transaction Publishers, 2007).
37. Grover Krantz, M. Halpin, and M. M. Ames, eds. *Manlike Monsters on Trail: early records and modern evidence.* (Vancouver: University of British Columbia Press, 1980).
38. Title page. *Cryptozoology*, 1 (1982).
39. Bernard Heuvelmans. "What is Cryptozoology?" *Cryptozoology* 1 (Winter 1982): 1–12.
40. Ibid., 1.
41. See materials in GSK, box 6, NAA.
42. Collection of notes in folder 0200, box 2 & 3, NAA.
43. For the Kennewick Man case and NAGPRA see; James Chatters, *Ancient Encounters: Kennewick Man and the First Americans* (New York: Simon & Schuster, 2002) and David Hurst Thomas, *Skull Wars: Kennewick Man, Archaeology, And the Battle for Native American Identity* (New York: Basic Books, 2001).
44. Don Sampson. *Ancient One/Kennewick Man*, November 21, 1997. Council of Federated Tribes of the Umatilla Indian Reservation, Pendleton, Oregon. Web site.
45. David Carkhuff. "Bigfoot on Congress Street: International Cryptozoology Museum due to open in Arts District Nov. 1," *Portland Daily Sun* (October, 2009). E-mail conversation with Loren Coleman, November 23, 2008.
46. For the Center for Fortean Zoology see cfz.org.uk.
47. See Nick Redfern. *Memoirs of a Monster Hunter: a Five-Year Journey in Search of the Unknown* (Franklin Lakes, NJ: Career Press, 2007).
48. BFRO web page bfro.net.
49. For the Southern Oregon group see the British Center for Bigfoot Research online. For Texas see, texasbigfoot.org.
50. This information gathered by the author through a survey in 2008 of a number of anomalous primate research organization members from across North America. This survey is illustrative only and should not be taken as definitive.

51. Richard Greenwell to GSK, 22 January 1985, folder 0342, box 7, NAA.

52. Police report #78-90469, December 20, 1975, City of Portland, Oregon, Police Department.

53. AAG Jennifer Hubbard Geller, Washington Attorney General's Office Memorandum, February 26, 1996, folder 0343, box 7, NAA.

54. René Dahinden to Richard Greenwell, 10 November 1982, folder 0342, box 7, NAA.

55. Ed Penhale. Seattle Post Intelligencer, 12/2/1985, D2.

56. René Dahinden to John Bodley. December 14, 1995, folder 0341, box 7, NAA.

57. René Dahinden to GSK. September 17, 1997, folder 0341, box 7, NAA.

58. This information comes from an online site, Bigfoot.org, quoting musician turned Bigfoot researcher Henry Franzoni.

59. Undated rough draft reply to Dmitri Bayanov's letter to GSK dated May 2, 1991.

60. GSK to René Dahinden. May 14, 1975, folder 0432, box 7, NAA.

61. GSK to René Dahinden. April 2, 1976, folder 0342, box 7, NAA.

62. John Green to John ? 7/8/1976, folder 0335, box 3, NAA.

63. René Dahinden to GSK. 5/12/1975, folder 0334, box 3, NAA.

64. René Dahinden to GSK. 5/26/1975, folder 0334, box 3, NAA.

65. Peter Byrne to GSK. 6/19/1975, folder 0334, box 3, NAA.

66. GSK to Robert Gottlieb. 6/24/1975, folder 0334, box 3, NAA.

67. GSK, Letters to the Editor column, True 56:461 (October, 1975).

68. René Dahinden to John Bodley. 14 December 1995, folder 0341, box 7, NAA.

69. René Dahinden to Mike Quast. January 15, 1993, folder 0343, box 7, NAA.

70. Richard Greenwell to René Dahinden. September 18, 1985, folder 0342, box 7, NAA.

71. Alan Campbell to GSK, 11 June 1998, folder 0341, box 7, NAA.

72. René Dahinden to GSK, 23 May 2000, folder 0341, box 7, NAA.

73. René Dahinden to Hancock House Publishers. July 14, 1999, folder 0341, box 7, NAA.

74. AAG Jennifer Hubbard Geller, Washington Attorney General's Office Memorandum, February 26, 1996, folder 0343, box 7, NAA.

75. René Dahinden to GSK. 1982, folder 0342, box 7, NAA.

76. GSK to René Dahinden. July 12, 1976, folder 0342, box 7, NAA.

77. B. A. Nugent to Jon Beckjord. October 3, 1978, WSU.

78. GSK to René Dahinden. August 7, 1987, folder 0342, box 7, NAA.

79. René Dahinden to GSK. April 6, 1982, folder 0342, box 7, NAA.

80. "New Pictures of Bigfoot," The Sun (February 13, 1996): 2–3.

81. See the BFRO web site.

82. Quoted in Mark Hume, "Trail Ends for Bigfoot's Biggest fan," The National Post (4/27/2001).

83. GSK, *Bigfoot/Sasquatch,* p. 86.
84. Benjamin Orlove to GSK, 2/15/2001, folder 0247, box 4, NAA.
85. Review sent by Benjamin Orlove to GSK, 9/4/2001, folder 0247, box 4, NAA.
86. Loren Coleman. Obituary of Grover Krantz, *Cryptozoologist* web site (March 2002).
87. Interview with John Bodley, February 2010.
88. Michael A. Woodley, Darren Naish, and Hugh P. Shanahan. "How many Extant Pinniped Species Remain to be Discovered?" *History of Biology* 20:4 (December 2008): 225–35.
89. William Straus. "Abominable Snowman," *Science* 127:3303 (April 18, 1958): 882–84, 883.
90. Quoted in Tom Paulson, "A Student of Sasquatch, Professor Grover Krantz Dies," *Seattle Post-Intelligencer Reporter* (February 18, 2002).
91. Interview with Richard Freeman, February 13, 2010.
92. Ivan Sanderson, "Abominable Snowmen are Here!" *True* 42: 294 (November 1961): 40–41, 86–92.
93. William Charles Osman-Hill. "The Abominable Snowmen: the present position," *Oryx* VI (1961): 86–98.
94. George Agogino. "An Overview of the Yeti-Sasquatch Investigations and Some Thoughts on Their Outcome," *Anthropological Journal of Canada* 16:2 (1978): 11–13.

## Notes on Sources and Monster Historiography

1. Those at the more valuable end of the spectrum include John Green, *Year of the Sasquatch: Encounters with Bigfoot from California to Canada* (Agassiz, B.C: Cheam Publishing, Ltd., 1970); Loren Coleman, *Bigfoot! The True Story of Apes in America* (New York: Paraview Books, 2003); Christopher Murphy, *Meet the Sasquatch* (Surrey, B.C.: Canada:: Hancock House Pub. Ltd., 2004); and Jonathan Downes, *Monster Hunter: In Search of Unknown Beasts at Home and Abroad* (Devon, U.K: Center for Fortean Zoology, 2004).
2. Chad Arment. *Cryptozoology: Science and Speculation* (Coachwhip Pub.: Landisville, PA, 2004).
3. Examples are Gorden Strasenburgh, "On *Paranthropus* and 'relic hominids,'" *Current Anthropology* 16 (1975): 486–87; Turhon A. Murad, "Teaching Anthropology and Critical Thinking with the Question "Is there something Big Afoot?" *Current Anthropology* 29:5 (Dec., 1988): 787–89; Bacil F. Kirtley, "Unknown Hominids and New World Legends," *Western Folklore* 23:2 (April, 1964:77–90); Linda Milligan, "The 'Truth' about the Bigfoot Legend," *Western Folklore* 49:1 (January, 1990): 83–98; Phillips Stevens, Jr., "'New' Legends: some perspectives from anthropology," *Western Folklore* 49:1 (Jan., 1990):121–33; and Peter Dendle, "Cryptozoology in the Medieval and Modern Worlds," *Folklore* 117 (August 2006): 190–206.

4. Jeffrey Jerome Cohen. *Monster Theory: Reading Culture* (University of Minnesota Press, 1996).

5. G. W. Gill, "Population clines of the North American sasquatch as evidenced by track lengths and estimated statures," in *Manlike Monsters on Trail: early records and modern evidence.* M. Halpin and M. M. Ames eds., (Vancouver: University of British Columbia Press, 1980); Jeff Meldrum *Sasquatch: Legend Meets Science* (Forge Books: New York, 2006), John Bindernagel. *The Discovery of Sasquatch* (Courtney, B.C.: Beachcomber Books, 2010).

6. Greg Long. *The Making of Bigfoot: the inside story* (Amherst, NY: Prometheus Books, 2004); Michael Mcleod. *Anatomy of a Beast: Obsession and Myth on the Trail of Bigfoot* (University of California Press, 2009); and Joshua Buhs, *Bigfoot: The Life and Times of a Legend* (University of Chicago Press, 2009).

7. See Harriet Ritvo. *The Platypus and the Mermaid and other Figments of the Classifying Imagination* (Harvard University Press: Cambridge, MA, 1997); and Lorraine Daston and Katherine Park. *Wonders and the Order of Nature 1150–1750* (New York: Zone Books, 1998).

8. Daniel Perez, *Big Footnotes* (D. Perez Pub., 1988).

# Bibliography

## Unattributed Articles

"'Ape man' escapes FBI," *Sunday Times*, London (April 27, 1969).

"Apemen May Still Be Living: Reports of Mongolian Discoveries," *Manchester Guardian* (July 12, 1958).

"Bigfoot Photo Baffles Experts!" *Weekly World News* (April 29, 1980).

"Cold freezes hounds off humanoid's trail," *Montana Standard* (December 7, 1969).

"Eugenics Conference Opens Here Today," *New York Times* (August 21, 1932): 15.

"Everest Headman Gives a Yeti Call in London," *Evening Standard* (December 12, 1960).

"Fights Gini Appointment," *New York Times* (February 14, 1935): 4.

"Moscow Suspicious of Hillary," *New York Times* (September 18, 1960): 45.

"New Pictures of Bigfoot," *The Sun* (February 13, 1996): 2–3.

"Obituary of Bernard Heuvelmans," *Fortean Times* 153 (December 2001).

"Papers Relating to the Himalaya and Mount Everest," *Proceedings of the Geographical Society of London* IX (April–May 1857): 345–51.

"Pooh-Pooh to Snowman," *New York Times* (January 10, 1960): 23.

"Professor: Track Down, Kill a Bigfoot," *Salt Lake Tribune* (January 22, 1996).

"Russians Doubt 'Snowman,'" *New York Times* (February 3, 1958): 4.

"Six men-with nylon ropes-to attack Everest," *News Chronicle* (December 18, 1945).

"Soviet Scientists Trail 'Snowman,'" *New York Times* (November 16, 1958): 122.

"Soviet See Espionage in US Snowman Hunt," *New York Times* (April 27, 1957): 8.

"Texan Balked in Nepal Hunting," *New York Times* (October 7, 1956):10.

"The 'Snowman' of the Pamirs," *Times*, London (January 16, 1958).

"The Iceman's magical mystery tour," *Sunday Times*, London ( September 28, 1969).

"U Lecturer from West Has Hunted 'Snowman,'" *Minneapolis Star*, (January 25, 1968).

Adams, Vincanne. *Tigers of the Snow: and other virtual Sherpas* (Princeton: Princeton University Press, 1995).

Agogino, George. "An Overview of the Yeti-Sasquatch Investigations and Some Thoughts on Their Outcome," *Anthropological Journal of Canada* 16:2 (1978): 11–13.

Alberti, Samuel J. M. M. "Amateurs and Professionals in One Country: Biology and Natural History in Late Victorian Yorkshire," *Journal of the History of Biology* 34 (2001): 115–47.

Alexander, Shana. "More Monsters Please!" *Life* (December 8, 1967): 308.

Allen, David Elliston. *The Naturalist in Britain: A Social History* (London: Penguin Books, 1976).

Anderson, David C. "Stalking the Sasquatch," *New York Times* (January 20, 1974): 231.

Anonymous. *Denver News* (May 5, 1988): 1A & 10A.

Anonymous. *Seattle Times* (February 13, 1972).

Arment, Chad. *Cryptozoology: Science and Speculation* (Landisville, PA: Coachwhip Pub., 2004).

Arment, Chad. *The Historical Bigfoot* (Landisville, PA: Coachwhip Pub., 2006).

Assouline, Pierre. *Hergé: the Man Who Created Tin Tin* (Oxford: Oxford University Press, 2009).

Barrow Jr., Mark V. *A Passion for Birds: American Ornithology after Audubon* (Princeton: Princeton University Press, 1998).

Bartholomew, Paul et al. *Monsters of the Northwoods* (Utica, NY: North Country Books, 1992).

Bartholomew, Robert and Brian Regal. "From Wild Man to Monster: the Historical Evolution of Bigfoot in New York State," *Voices: the Journal of New York Folklore* 35: 3–4 (Fall–Winter 2009): 13–15.

Bartlett, Kay. "Bigfoot Hunters Don't Get Along," *Times-Union and Journal*, Jacksonville, Florida (January 28, 1979): A1–A4.

Barton, Ruth. "Huxley, Lubbock, and a Half Dozen Others: Professionals and Gentleman in the Formation of the X Club, 1851–1864," *Isis* 89 (1998): 410–44.

Bayanov, Dmitri. "Letters in Response to 'Bigfoot Believers,'" *Bigfoot Information Project* website (August 8, 2004).

Bayanov, Dmitri. *Bigfoot Research: the Russian View* (Moscow: Crypto-Logos, 2007).

Bayanov, Dmitri. *In the Footsteps of the Russian Snowman* (Moscow: Crypto-Logos, 1996).

Beehler, Bruce, Roger Pasquier, and Warren King. "In Memoriam: S. Dillon Ripley, 1913–2001," *The Auk* (October 2002).

Bishop, Peter. *The Myth of Shangri-La: Tibet, Travel Writing and the Western Creation of Sacred Landscape* (Berkeley, CA: University of California Press, 1989).

Blancou, Lucien. *Géographie Cynégétique du Monde* (Paris: Presses Universitaires de France, 1959).

Bledstein, Burton J. *The Culture of Professionalism: The Middle Class and the Development of Higher Education in America* (New York: Norton, 1976).

Bodley, John. "Sasquatch Footprints: Can Dermal Ridges be Faked?" *Northwest Science* 62:3 (1988): 129–30.

Bohme, Gernot. "Alternatives in Science—alternatives to Science," in *Counter-Movements in the Sciences*, H. Rose, ed. *Sociology of Science Yearbook*, volume 3 (Boston: D. Reidel, 1979):105–25.

Bovey, Alixe. *Monsters and Grotesques in Medieval Manuscripts* (The British Library, London: 2002).

Bowler, Peter. *Science for All: the Popularization of Science in Early Twentieth-Century Britain* (Chicago: University of Chicago Press, 2009).

Bowman, Matthew. "A Mormon Bigfoot: David Patten's Cain and the Conception of Evil in LDS Folklore," *Journal of Mormon History* 33:3 (Fall 2007): 62–82.

Brink-Roby, H. "*Siren canora*: the Mermaid and the Mythical in Late Nineteenth-Century Science," *Archives of Natural History* 35:1 (2008): 1–14.

Bruce, Robert V. *The Launching of Modern American Science, 1846–1876* (Ithaca, NY: Cornell University Press, 1987).

Buhs, Joshua Blu. Bigfoot: Life and Times of a Legend (Chicago: University of Chicago Press, 2009).

Burns, J. W. and C. V. Tench. "The Hairy Giants of British Columbia," *Wide World Magazine* (January 1940).

Busse, Phil. "Looking for Mr. Bigfoot: The Western Bigfoot Society and the Eternal Search for Truth," *Portland Mercury* (September 14, 2000).

Byrne, Peter. *The Search for Bigfoot: Monster, Myth or Man?* (Washington, D.C.: Acropolis Books, 1975).

Carkhuff, David. "Bigfoot on Congress Street: International Cryptozoology Museum due to open in Arts District Nov. 1," *Portland Daily Sun* (October 2009).

Carranco, Lynwood. "Three Legends of Northwestern California," *Western Folklore* 22:3 (July 1963): 179–85.

Casey, Phil. "Strange Iceman Tale: A New Race or an Old Hoax?" *Washington Post* (March 27, 1969).

Challinor, David. "In Honor of Dillon Ripley," *Proceedings of the American Philosophical Society* 147:3 (September 2003).

Chatters, James. *Ancient Encounters: Kennewick Man and the First Americans* (New York: Simon & Schuster, 2002).

Ciochon, Russell, John Olsen, and Jamie James. *Other Origins: The Search for the Giant Ape in Human Prehistory* (New York: Bantam, 1990).

Clark, Jerome and Loren Coleman, *Cryptozoology A-Z* (New York: Fireside, 1999).

Cohen, Jeffrey Jerome. *Monster Theory: Reading Culture* (Minneapolis, MN: University of Minnesota Press, 1996).

Coil Suchy, Linda. *Who's watching You? An exploration of the Bigfoot phenomenon in the Pacific Northwest* (Blaine, WA: Hancock House, 2009).

Coleman, Loren. "The Dalai Lama, Slick Denials and the CIA," in *Popular Alienation: A Steamshovel Press Anthology* (Kempton, IL: IllumiNet Press, 1995).

Coleman, Loren. *Bigfoot! The true story of apes in America* (New York: Paraview Books, 2003).

Coleman, Loren. Obituary of Grover Krantz, *Cryptozoologist* web site (March 2002).

Coon, Carleton S. *A North Africa Story: The Anthropologist as OSS Agent: 1941–1943* (Massachusetts: : Gambit, 1980).

Coon, Carleton S. *The Story of Man: from the first human to primitive culture and beyond* (New York: Alfred A. Knopf, 1954).

Coon, Carleton S., *Adventures and Discoveries: The Autobiography of Carleton S. Coon* (Prentice Hall, 1981).

Coon, Carleton. "Why There has to be a Sasquatch," in Markotic, Vladimir and Grover Krantz, eds. *The Sasquatch and Other Unknown Hominoids* (Calgary: Western Publishers, 1984).

Coon, Carleton. Review of "Sapienization and Speech," *Current Anthropology* (April 23, 1980).

Coon. Carleton. *The Races of Europe* (New York: Macmillan, 1939) and *Origin of the Races* (New York: Knopf, 1962).

Corbey, Raymond. *The Metaphysics of Apes: Negotiating the Animal-Human Boundary* (Cambridge: University of Cambridge Press, 2005).

Daegling, David J. *Bigfoot Exposed: An Anthropologist Examines America's Enduring Legend* (Walnut Creek, CA: Altimira Press, 2004).

Dahinden René with Don Hunter. *Sasquatch* (Toronto: McClelland and Stewart, 1973).

Daniels, George H. "The Process of Professionalization in American Science: The Emergent Period, 1820–1860," *Isis*, 1967, 58: 151–66.

Daniels, George H. *American Science in the Age of Jackson* (New York: Columbia Univ. Press, 1968).

Daston, Lorraine and Katherine Park. *Wonders and the Order of Nature 1150–1750* (New York: Zone Books, 1998).

David L. Snellgrove. *The Cultural History of Tibet* (Orchid Press: 2006).

De Sarre, François. "Essai sur le Statut Phylogenique des Hominoides Fossiles et Recents: la point devue de la theorie de la bipedie," *Bipedia* 5.1 (Septembre 1990).

Dendle, Peter. "Cryptozoology in the Medieval and Modern World," *Folklore* 117 (August 2006):190–206.

Dennett, Michael and Daniel Loxton, "Some reasons for Caution about the Bigfoot Film Expose," *Skeptical Inquirer* (Jan–Feb 2005).

Dennett, Michael R. "Bigfoot evidence: are these tracks real?" *Skeptical Inquirer* (September 22, 1994).

Desmond, Adrian. "Redefining the X Axis: "Professionals," "Amateurs" and the Making of Mid-Victorian Biology—A Progress Report," *Journal of the History of Biology* 34 (2001): 3–50.

Desmond, Adrian. *Huxley: From Devil's Disciple to Evolution's High Priest* (Reading, MA: Addison-Wesley, 1997).

Dobzhansky, Theodosius. *Mankind Evolving: the Evolution of the Human Species* (New Haven: Yale University Press, 1962).

Dobzhansky, Theodosius. "Genetic Entities in Hominid Evolution," in Sherwood Washburn, ed. *Classification and Human Evolution* (Piscataway, NJ: Aldine Pub. Co., 1963).

Dornin, Rusty. "Don't Believe in Aliens? Visit San Francisco's UFO Museum," *CNN Interactive* (April 19, 1997).

Douglas, Mac. "The Snowman," *Everybody's Magazine* (December 6, 1958).

Downes, Jonathan. *Monster Hunter: In Search of Unknown Beast at Home and Abroad* (Devon, U.K: Center for Fortean Zoology, 2004).

Downing, Bob. "Chief Hunter of Bigfoot Calls it Quits," *San Francisco Examiner and Chronicle* (September 30, 1979): B:5.

Duffus, R. L. "The Plan of Attack is on War Itself," *New York Times* (January 11, 1959): br3.

Dupree A. Hunter. *Science in the Federal Government: A History of Policies and Activities to 1940* (Cambridge, MA: Belknap Press, 1957).

Elman, Robert. *First in the Field: America's Pioneering Naturalists* (New York: Mason/Charter, 1977).

Endersby, Jim. *Imperial Nature: Joseph Hooker and the Practices of Victorian Science* (Chicago: University of Chicago Press, 2008).

Evans, Howard Ensign. *Pioneer Naturalists: The Discovery and Naming of North American Plants and Animals* (New York: Henry Holt & Co., 1993).

Faulkner, Alex. "US Film of Abominable Snowman," *The Daily Telegraph*, London (November 22, 1967).

Favero, Giovanni. *Il Fascismo Razionale: Corrado Gini fra Scienza e Politica* (Rome: Carocci, 2006).

Fitter, Richard. "Mask Provides New Clues to Snowman," *The Observer* (October 23, 1960).

Fort, Charles Hoy. *Book of the Damned* (1919).

Forth, Gregory. "Flores after floresiensis Implications of local reaction to recent paleoanthropological discoveries on an eastern Indonesian island," *Bijdragen tot de Taal-, Land- en Volkenkunde* (BKI) 162-2/3 (2006): 336–49.

Forth, Gregory. "Hominids, hairy hominoids and the science of humanity," *Anthropology Today* 21:3 (June 2005): 13–18.

Forth, Gregory. *Images of the Wildman in Southeast Asia: An Anthropological Perspective* (New York: Routledge, 2008).

Friedman, John Block. *The Monstrous Races in Medieval Art and Thought* (Cambridge, MA: Harvard University Press, 1981).

Genzoli, Andrew. "RFD," *Humboldt Times* (October 1, 1961).

Gill, G.W. "Population clines of the North American Sasquatch as evidenced by track lengths and estimated statures," in *Manlike Monsters on Trail: early records and modern evidence.* M. Halpin and M. M. Ames eds., (Vancouver: University of British Columbia Press, 1980).

Gillette, Aaron. *Eugenics and the Nature-Nurture Debate in the Twentieth-Century* (New York: Palgrave-Macmillan, 2007).

Gillette, Aaron. *Racial Theories in Fascist Italy* (New York: Routledge, 2002).

Gini, Corrado. "The Scientific Basis of Fascism," *Popular Science Quarterly* 42:1 (March 1927): 99–115.

Gladwin, Harold Sterling. *Men Out of Asia: An Exciting Picture of the Early Origins of Early American Civilization* (New York: McGraw Hill, 1947).

Green, John. "What is a Sasquatch?" in *Manlike Monsters on Trail: early records and modern evidence.* M. Halpin and M. M. Ames eds., (Vancouver: University of British Columbia Press, 1980).

Green, John. *On the Track of Sasquatch* (Agassiz, BC: Cheam Pub. Inc., 1968).

Green, John. *Sasquatch: the Apes Among Us* ( Blaine, WA: Hancock House Publishers, 2006).

Green, John. *Year of the Sasquatch* (Agassiz, BC: Cheam Pub. Ltd., 1970).

Green, Mary and Janice Coy. *Fifty Years with Bigfoot* (self-published, 2002).

Greene John C. *American Science in the Age of Jefferson* (Ames: Iowa State University Press, 1984).

Greenwell, Richard and James E. King, "Attitudes of Physical Anthropologists towards Reports of Bigfoot and Nessie," *Current Anthropology* 22:1 (February 1981): 79–80.

Grumley. Michael. *There Were Giants in the Earth* (Garden City, NY: Doubleday & Co., Inc., 1974).

Hall, Mark A. "Biography of Ivan Sanderson," *Wonders* (December 1992): 65–67.

Hanafi, Zakiya. *Monster in the Machine: Magic, Medicine, and the Marvelous in the Time of the Scientific Revolution* (Durham, NC: Duke University Press, 2000).

Hansen, Frank. "I killed the Ape-Man Creature of Whiteface," *Saga* (July 1970).

Hatch, Nathan O. ed. *The Professions in American History* (Notre Dame, IN: Univ. Notre Dame Press, 1988).

Heaney, Michael. "The Mongolian Almas: a Historical Reevaluation of the Sighting by Baradiin," *Cryptozoology* 2 (1983): 40–52.

Heobell, A. Adamson. *Man in the Primitive World* (New York: McGraw-Hill, 1949).

Heuvelmans, Bernard and Boris Porchnev. *L'Homme De Néanderthal est Toujours Vivant* (Librairie Plon, 1974).

Heuvelmans, Bernard. "L'Homme des Cavernes a-t-il connu des Géants mesurant 3 à 4 mètres?" *Science et Avenir* 61 & 63 (Mai 1952).

Heuvelmans, Bernard. "Note Preliminaire sur un Specimen Conserve dans la glace d'une forme encore inconnu d'Hominide Vivian Homo Pongoides," *Bulletin Institute Royale des Sciences Naturelles de Belgique* 45:4 (1969): 1–24.

Heuvelmans, Bernard. "The Birth and Early History of Cryptozoology," *Cryptozoology* 3 (1984): 1–30.

Heuvelmans, Bernard. "What is Cryptozoology?" *Cryptozoology* 1 (Winter 1982): 1–12.

Heuvelmans, Bernard. *In the Wake of the Sea-Serpents* (New York: Hill & Wang, 1968).

Heuvelmans, Bernard. *On the Track of Unknown Animals* (New York: Hill & Wang, 1958).

Heuvelmans, Bernard. *Sur la Piste des Bêtes Ignorées* (Librarie Plon, 1955).

Hill, Bill. "Seeing the Sasquatch and the Tooth Fairy," *Lewiston Morning Tribune* (March 6, 1988).

Hillary, Sir Edmund. "The Scalp is not a Scalp at all," *Sunday Times* (January 15, 1961): 5.

Hindle, Brooke. The *Pursuit of* Science in Revolutionary America, *1735–1789* (Chapel Hill: University of North Carolina Press, 1956).

Holmfeld, John D. "From Amateurs to Professionals in American Science: The Controversy over the Proceedings of an 1853 Scientific Meeting," *Proceedings of the American Philosophical Society* 114:1 (February 1970): 22–36.

Hooton, Ernest. "Pessimist's Proposal," *Time* (March 30, 1936).

Isserman, Maurice and Stewart Weaver. *Fallen Giants: A History of Mountaineering from the Age of Empire to the Age of Extremes* (Yale University Press, 2008).

Izzard, Ralph. *An Innocent on Everest* (New York: EP Dutton & Co., 1954).

Izzard, Ralph. *Evening Standard* (December 30, 1960).

Izzard, Ralph. *The Abominable Snowman* (New York: Doubleday & Co., 1955).

Jackson Jr., John P. "In Ways Unaccademical": The Reception of Carleton S. Coon's *The Origin of the Races,*" *Journal of the History of Biology* 34 (2001): 247–85.

Jenkins, Alan C. *The Naturalists: Pioneers of Natural History* (New York: Mayflower Books, Inc., 1978).

Kirkpatric, Dick. "The Search for Bigfoot: Has a 150 Year-Old Legend Come to Life on this Film?" *National Wildlife* (April–May 1968): 43–47.

Kirtley, Bacil F. "Unknown Hominids and New World Legends," *Western Folklore* 23:2 (April 1964) 77–90.

Kisling, Vernon K. *Zoo and Aquarium History* (James Ellis Pub., 2000).

Klement, Jan. *The Creature: Personal Experiences with Bigfoot* (Allegheny Press: Elgin, PA, 2006).

Knoppers, Laura Lunger and Joan B. Landeseds., *Monstrous Bodies/Political Monstrosities in Early Modern Europe* (Ithaca, NY: Cornell University Press, 2004).

Kohler, Robert E. *All Creatures: Naturalists, Collectors, and Biodiversity, 1850–1950* (Princeton: Princeton University Press, 2006).

Korff, Kal K. "The Making of Bigfoot," *Fortean Times* 119 (February 2005): 34–39.

Krantz, Grover, "*Homo erectus* Brain Size by Subspecies," *Human Evolution* 10:2, 1995: 107–17.

Krantz, Grover, "Pithecanthropine Brain Size and its Cultural Consequences," *Man* 61, (May 1961): 85–87.

Krantz, Grover, *The Origins of Man* (University of Minnesota, UMI Dissertation Services, 1971), 13.

Krantz, Grover. "Additional Notes on Sasquatch Foot Anatomy," *North American Research Notes* 6:2 (Fall 1972).

Krantz, Grover. "Anatomy of a Sasquatch Foot," *North American Research Notes* 6:1 (Spring 1972): 91- 104.

Krantz, Grover. "Brain Size and Hunting Ability in Earliest Man," *Current Anthropology* 9:5, (December 1968): 450–51.

Krantz, Grover. "Review of *Human Variation: Races, Types, and Ethnic Groups*," by Stephen Molnar, *American Anthropologist* 85 (1983).

Krantz, Grover. "Sapienization and Speech," *Current Anthropology*. 21:6, (December 1980): 773–92.

Krantz, Grover. "Sasquatch Handprints," *North American Research Notes* 5:2, (Fall 1971): 145–51.

Krantz, Grover. "Sphenoidal Angle and Brain Size," *American Anthropologist* 64:3 (1962): 521–23.

Krantz, Grover. "The Populating of Western North America," *Method and Theory in California Archaeology* 1 (1977): 1–63.

Krantz, Grover. *Big Footprints: A Scientific Inquiry Into the Reality of Sasquatch* (Boulder CO: Johnson Books, 1992).

Krantz, Grover. *Climatic Races and Descent Groups* (North Quincy, MA.: Christopher Publishing House, 1980).

Krantz, Grover. *East Washingtonian*, Pomeroy, WA. August 16, 1971.

Krantz, Grover. *Only a Dog*, (Wheat Ridge, CO: Hoflin Pub., 1998).

Krantz, Grover. *The Process of Human Evolution* (Cambridge, MA: Schenkman, 1981).

Lester, Shane. *Clan of Cain: the Genesis of Bigfoot* (self published, 2001).

Ley, Willy. "Do Prehistoric Monsters Still Exist?" *Mechanix Illustrated* (February 1949):80–144.

Ley, Willy. *Exotic Zoology* (New York: Viking Press, 1959).

Ley. Willy. *Salamanders and Other Wonders* (New York: Viking Press, 1955).

Loftus, Bill. "Professor's Sasquatch Hunt a Private Matter," *Lewiston Morning Tribune* (April 4, 1988).

Long, Greg. *The Making of Bigfoot: the inside story* (Amherst, NY: Prometheus Books, 2004).

Lucier, Paul. "The Professional and the Scientist in Nineteenth-Century America," *Isis* 100:4 (December 2009): 699–732.

M. H. Day, "In Memoriam," *Journal of Anatomy* 159 (1988): 227–29.

MacGregor, Greg. "World is Asked to Accept China," *New York Times* (January 16, 1961): 3.

Markotic, Vladimir and Grover Krantz eds., *The Sasquatch and Other Unknown Hominoids* (Calgary: Western Publishers, 1984).

Mayfield, Harold F. "The Amateur in Ornithology," *The Auk* 96 (January 1979): 168–71.

Mayor, Adrienne. *Fossil Legends and the First Americas* (Princeton: Princeton University Press, 2007).

McCartney, Eugene S., "Modern Analogs to Ancient Tales of Monstrous Races," *Classical Philology* 36:4 (October 1941): 390–94.

McCown, Theodore. *Joint Expedition of the British School of Archaeology in Jerusalem and the American School of Prehistoric Research (1929–1934): The Stone Age of Mount Carmel* (Oxford: Clarendon Press, 1937–1939).

Mcleod, Michael. *Anatomy of a Beast: Obsession and Myth on the Trail of Bigfoot* (Berkeley, CA: University of California Press, 2009).

Meldrum, Jeffrey and Trent Stevens. *Evolution and Mormonism: a quest for understanding* (Signature Books, 2001).

Meldrum, Jeffrey. "Who are the Children of Lehi," *FARMS Review* (2001).

Meldrum, Jeffrey. *Sasquatch: Legend Meets Science* (New York: Forge Books, 2006).

Meyer, Amy and Margaret Pritchard, eds. *Empire's Nature: Mark Catesby's New World Vision* (ChapellHill, NC: University of North Carolina Press, 1998).

Miles, Ray. *King of the Wildcatters: the Life and Times of Tom Slick, 1883–1930* (College Station, TX: Texas A&M University Press, 1996).

Milligan, Linda. "The "Truth" about the Bigfoot Legend," *Western Folklore* 49:1 (January, 1990): 83–98.

Monaghan, Peter. "Cryptozoologists defy other scientists' skepticism to stalk beasts found in legend, art, and history," *The Chronicle of Higher Education* (February 10, 1993): A7–A9.

Montagna, William. "From the Director's Desk," *Primate News* (September 1976): 7–9.

Murad, Turhon A. "Teaching Anthropology and Critical Thinking with the Question 'Is there something Big Afoot?'" *Current Anthropology* 29:5 (December 1988): 787–89.

Murphy, Christopher. *Bigfoot Film Journal* (Blane, WA: Hancock House Publishers (ebook): 2008): 75.

Murphy, Christopher. *Meet the Sasquatch* (Blaine, WA: Hancock House Pub., Ltd., 2004).

Napier, John. *Bigfoot: the Yeti and Sasquatch in Myth and Reality* (New York: E.P. Dutton & Co., 1972).

*National Cyclopedia of American Biography* vol. 57 (Clifton, NJ: James T. White & Company, 1977): 192–94.

Anonymous, *New York Times* (December 20, 1953).

Nixon Cooke, Katherine. *Tom Slick: Mystery Hunter* (Bracy, VA: Paraview Inc., 1995).

Noll, Richard. "Interview with Dr. Grover Krantz," *Bigfoot Encounters.com* (July 1, 2001).

Norman, Eric. *The Abominable Snowman* (New York: Award Books, 1969): 22

O'Brien, Michael J. and R. Lee Lyman. *Applying Evolutionary Archaeology: A Systematic Approach* (Springer, 2000).

Oakley, Don and John Lane. "Earth, Stars and Man . . . Ape Men and Giants," *Yakima Daily Republic* (November 2, 1960).

Ogilvie, Brian W. *Science of Describing: Natural History in Renaissance Europe* (Chicago: University of Chicago Press, 2006).

Oosterzee, Penny van. *The Story of Peking Man* (Cambridge, MA: Perseus Publishing, 2000).

Ortner, Everett. "Do 'Extinct' animals still survive?" *Popular Science Monthly* (1959).

Ortner, Sherry B. *High Religion: A Cultural and Political History of Sherpa Buddhism* (Princeton: Princeton University Press, 1989).

Osman-Hill, William Charles. "Man's Relation to the Apes," *Man* 50 (December 1950): 161–62.

Osman-Hill, William Charles. "Note on the Nomenclature of Certain Hominidae," *American Journal of Physical Anthropology* 9:3 (September 1951): 1–3.

Osman-Hill, William Charles. "Nittaewo: An Unsolved Problem from Ceylon," *Loris: A Journal of Ceylon Wildlife* 4:1 (1945): 251–62.

Osman-Hill, William Charles. "The Abominable Snowmen: the present position," *Oryx* VI (1961): 86–98.

Oudemans, A. C. *The Great Sea Serpent: An historical and critical treatise* (Leiden: E. J. Brill; London: Luzac & co., 1892).

Patterson, Roger and Christopher Murphy. *The Bigfoot Film Controversy* (Blaine, WA: Hancock House, 2005).

Patterson, Roger. *Do Abominable Snowmen of America Really Exist?* (Yakima, WA: Franklin Press, 1966).

Paulson, Tom. "A Student of Sasquatch, Professor Grover Krantz Dies," *Seattle Post-Intelligencer Reporter* (February 18, 2002).

Peissel, Michael. *Tiger for Breakfast: The Story of Boris of Xathmandu* (New York: EP Dutton Co., 1966): 233.

Penhale, Ed. *Seattle Post Intelligencer*, (December 2, 1985), D2.

Perez, Daniel. *Big Footnotes* (D. Perez Pub., 1988).

Perez, Daniel. *Bigfoot at Bluff Creek* (Norwalk, CA: Center for Bigfoot Studies, 2003).

Perez, Daniel. Center for Bigfoot Studies, CA, "Review of *Bigfoot/Sasquatch Evidence*," 1999.

Pincher, Chapman. "Hillary Leads New Snowman Hunt," *Daily Express* (May 6, 1960).

Pocock, Chris. *Dragon Lady: History of the U-2 Spy Plane* (Airlife, 1989).

Porshnev, Boris. "The Problem of Relic Paleoanthropus," *Soviet Ethnography* 2 (1969): 115-30.

Porshnev, Boris. "The Troglodytidae and the Hominidae in the Taxonomy and evolution of higher primates," *Current Anthropology* 15:449, (1974): 450.

Porshnev, Boris. "Troglodytidy i gominidy v sistematike i evolutsii vysshikh primatov," *Doklady Akademii Nauk SSSR*, 188:1 (1969).

Porshnev, Boris. *Leninist Theory of Revolution and Social Psychology* (Moscow: Novosti Press agency, nd).

Porshnev, Boris. *Social Psychology and History* Ivan Savin, trans., Vic Schneierson, ed. (Moscow: Progress Publishers, 1970).

Powell, Thom. *The Locals: A Contemporary Investigation of the Bigfoot/ Sasquatch Phenomenon* (Blaine, WA: Hancock House, 2003).

Prince-Hughes, Dawn. *The Archetype of the Ape-man* (Dissertation.com.: 2001).

Putnam, Carleton. *Race and Reason: a Yankee View* (Washington, D.C: Public Affairs Press, 1961).

Rainger, Ronald, *An Agenda for Antiquity* (Tuscaloosa, AL: University of Alabama Press, 1991).

Redfern, Nick. *Man-Monkey: in search of the British Bigfoot* (North Devon, U.K: Center for Fortean Zoology Press, 2007).

Redfern, Nick. *Memoirs of a Monster Hunter: a Five-Year Journey in Search of the Unknown* (Franklin, NJ: Career Books, 2007 ).

Reece, Dr. Gregory L. *Weird Science and Bizarre Beliefs* (London: I. B. Tauris, 2009).

Regal, Brian. "Amateur Versus Professional: The Search for Bigfoot," *Endeavour* 32:2 (June 2008): 53-57.

Regal, Brian. "Entering Dubious Realms: Grover Krantz, Science and Sasquatch," *Annals of Science* 66:1 (January 2009).

Regal, Brian. *Henry Fairfield Osborn: Race and the Search for the Origins of Man* (London: Ashgate, 2002).

Regal, Brian. *Human Evolution: A Guide to the Debates* (Santa Barbara: ABC-CLIO, 2004).

Reingold, Nathan. *Science, American Style* (New Brunswick, NJ: Rutgers Univ. Press, 1991).

Rensberger, Boyce. "Is it Bigfoot, or Can it be Just a Hoax?" *New York Times* (June 30, 1976): 78.

Rinčen, Yöngsiyebü. "L'Heritage Scientifique du Prof. Dr. Zamcarno," *Central Asiatic Journal* 4:3 (1959): 199-206.

Rinchen, P. R. "Almas Still Exist in Mongolia," *Genus* 20 (1964): 188-92.

Ritvo, Harriet. *The Platypus and the Mermaid and other Figments of the Classifying Imagination* (Cambridge: Harvard University Press, 1997).

Ruane, Michael E. "Written in Bone, exhibit," *Washington Post* (April 11, 2009): CO1.

Salim,Ali, B. Biwas, Dillion Ripley, and A. K. Gosh. *The Birds of Bhutan* (Zoological Survey of India: 2002).

Sanderson, Ivan "More Evidence that Bigfoot Exists," *Argosy* (April 1, 1968).

Sanderson, Ivan "The Missing Link," *Argosy* 368:5 (May, 1969): 23–31.

Sanderson, Ivan. "Abominable Snowmen are Here!" *True* 42:294 (November 1961): 40–41, 86–92.

Sanderson, Ivan. "First Photos of Bigfoot: California's 'Abominable Snowman,'" *Argosy* 29 (February 1968).

Sanderson, Ivan. "More about the Abominable Snowman," *Fantastic Universe* 2:6 (October 1959): 58–64.

Sanderson, Ivan. "Preliminary Description of the External Morphology of What Appeared to be a Fresh Corpse of a Hitherto Unknown Form of Living Hominid," *Genus* XXV: 1–4 (1969): 249–84.

Sanderson, Ivan. "Some Preliminary Notes on Traditions of Submen in Arctic and Subarctic North America," *Genus* XIX: 1–4 (1963): 145–62.

Sanderson, Ivan. "The Patterson Affair," *Pursuit* (June 1968).

Sanderson, Ivan. "The Strange Story of America's Abominable Snowman," *True* (December 1959).

Sanderson, Ivan. "The Wudewása or Hairy Primitives on in Ancient Europe," *Genus* XVIII: 1–4 (1962): 109–27.

Sanderson, Ivan. "There Could Be Dinosaurs," *Saturday Evening Post* (January 3, 1948).

Sanderson, Ivan. *Abominable Snowmen: Legend Come to Life* (Philadelphia: Chilton Co., 1961): 20.

Sanderson, Ivan. *Invisible Residents* (New York: Avon Books, 1973).

Sanderson, Ivan. *Ivan Sanderson's Anthology of Animal Tales* (New York: Alfred A. Knopf, 1946).

Sanderson, Ivan. *The Monkey Kingdom* (New York: Hanover House, 1957).

Sanderson, Ivan. *Things* (New York: Pyramid Books, 1967).

Sara Nelson, "Yeti Evidence is 'Convincing' says Wildlife Expert Sir David Attenborough," *Daily Mail On-Line* (March 1, 2009).

Saxon, Wolfgang. "Sherwood Washburn, Pioneer in Primate Studies Dies at 88," *New York Times* (April 9, 2000).

Schmeltzer, Michael. "Bigfoot Lives," *Washington Magazine* V (September–October 1998): 64–69.

Shackley, Myra. "Case for Neanderthal Survival," *Antiquity* LVI (1982): 31–41.

Shackley, Myra. *Neanderthal Man* (Hamden, CT: Archon, 1980).

Shackley, Myra. *Still Living? Yeti, Sasquatch and the Neanderthal Enigma* (New York: Thames & Hudson, 1983).

Shackley, Myra. *Wildmen: Yeti, Sasquatch and the Neanderthal Enigma* (Surrey, U.K.: Thames & Hudson, 1983).

Sheppard-Wolford, Sali. *Valley of the Skookum: Four Years of Encounters with Bigfoot* (Pine Winds Press, 2006).

Shipman, Pat. *The Evolution of Racism: Human Differences and the Use and Abuse of Science* (Cambridge, MA: Harvard University Press, 1994).

Shipton, Eric. *The Mount Everest Reconnaissance Expedition 1951* (London: Hodder and Stoughton, 1952).

Shipton, Eric. *The Six Mountain-Travel Books* (Diadem Books, Ltd., 1985).

Silverberg, Robert. *Mound Builders of Ancient America: The Archaeology of a Myth* (Greenwich, CT.: New York Graphic Society, 1968).

Simpson, George Gaylord. "The Beginnings of Vertebrate Paleontology in North America," *Proceedings of the American Philosophical Society* 86 (1943): 130–88.

Slate, B. Ann and Alan Berry. *Bigfoot* (New York: Bantam, 1976).

Slaughter, Thomas P. *The Natures of John and William Bartrum* (New York: Alfred A. Knopf, 1996),

Slick, Tom. "Yeti Expedition," *Explorer's Journal* (December 1958): 5–8.

Slick, Tom. *Permanent Peace: A Check and Balance Plan* (Prentice Hall, 1958).

Smith, Dwight G. and Gary Mangracopra, "What Readers Wrote In: secondary Bigfoot sources as given in the Letters-to-the-Editors column of the 1960s-1970s Men's Adventure Magazines," *North American BioFortean Review* 5:4 issue 13 (December 2003): 19–31.

*Smithsonian Torch* (October 1967): 2.

Somer, Lonnie, "New Signs of Sasquatch Activity in the Blue Mountains of Washington State," *Cryptozoology* 6 (1987): 65–70.

Soule, Gardner "The World's Most Mysterious Footprints," *Popular Science* (December 1952): 133–24.

Sprague, Roderick and Grover Krantz, eds. *The Scientist Looks at the Sasquatch*, (Moscow, ID: University Press of Idaho, 1977).

Sprague, Roderick. "Editorial," *North American Research Notes* 4:2 (Fall 1970): 127–28.

Stebbins, Robert A. "Serious Leisure: A Conceptual Statement," *Pacific Sociological Review* 25:2 (April 1982): 251–72.

Stebbins, Robert A. "The Amateur," *Pacific Sociological Review* 20:4 (October 1977): 588.

Stebbins, Robert A. *Amateurs: Margin Between Work and Leisure* (Beverley Hills, CA: Sage Publications, 1979).

Stebbins, Robert A. *Serious Leisure: a Perspective for our Time* (New Brunswick, NJ: Transaction Publishers, 2007).

Stevens Jr., Phillips. ""New" Legends: some perspectives from anthropology," *Western Folklore* 49:1 (January 1990): 121–33.

Stocking, George W,. Jr., ed. *Bones, Bodies, Behavior: Essays on Biological Anthropology* (Madison, WI: University of Wisconsin Press, 1988).

Stolyhwo, K, "Le crâne de Nowosiolka considéré com preuve de l'existence à l'époque historique de formes apparentées à *H. primigenius*," *Bulletin International de l'Académie des Sciences de Cracovie* (1908): 103–26.

Stonor, Charles *The Sherpa and the Snowman* (London: Hollis & Carter, 1955).

Strasenburgh, Gordon, "On *Paranthropus* and 'relic hominids,'" *Current Anthropology* 16 (1975): 486–87.

Straus Jr. William L. "Myth, Obsession, Quarry?" *Science* ns 136:3512 (April 20, 1962): 252.

Straus, William. "Abominable Snowman," *Science* 127:3303 (April 18, 1958): 882–84.

Stringer, Christopher and Robin McKie. *African Exodus: the Origins of Modern Humanity* (New York: Henry Holt, 1996).

Struit, Dirk Jan. *Yankee Science in the Making: Science and Engineering in New England from Colonial Times to the Civil War* (New York: Dover Publications, 1991).

Stump, Al. "The Man Who Hunts Bigfoot," *True* 56:456 (May 1975): 28–31, 74–77.

Sullivan, Robert. "Bigfoot," *Open Spaces Quarterly* 1:3 (1999).

Tattersall, Ian, Eric Delson, and John Van Couvering. *Encyclopedia of Human Evolution and Prehistory* (New York, Garland Publishing, 1988).

Taylor, Brian. "Amateurs, Professionals and the Knowledge of Archaeology," *British Journal of Sociology* 46:3 (September 1995): 499–508.

Tchernezky, W. "Nature of the 'Abominable Snowman: A New Form of Higher Anthropoid?'" *Manchester Guardian* (February 20, 1954).

Thomas, David Hurst. *Skull Wars: Kennewick Man, Archaeology, and the Battle For Native American Identity* (New York: Basic Books, 2001).

Thomas, Jill. "Russian Researcher Visits Overton County Tennessee," *Herald Citizen* (September 24, 2004).

Thorne, A.G., and M.H. Wolpoff. "Regional continuity in Australasian Pleistocene hominid evolution." *American Journal of Physical Anthropology* 55 (1981): 337–49.

Tschernezky, W. "A Reconstruction of the foot of the 'Abominable Snowman,'" *Nature* 186:4273 (May 7, 1960): 496–97.

Vertinsky, Patricia. "Physique as Destiny: William H. Sheldon, Barbara Honeyman Heath and the Struggle for Hegemony in the Science of Somatotyping," *CBMH/BCHM* 24:2 (2007): 291–316.

Vetter, Jeremy. "Cowboys, Scientists, and Fossils: The Field Site and Local Collaboration in the American West," *Isis* 99:2 (June 2008): 273–303.

Von Koenigswald, G. H. R. "Early Man: Facts and Fantasy," *The Journal of the Royal Anthropological Institute of Great Britain and Ireland* 94:2 (1964): 67–79.

Wangmo, Jamyang. *The Lawudo Lama: Stories of Reincarnation from the Mount Everest Region* (Somerville, MA: Wisdom Publications, 2005).

Washburn, Sherwood and Irven DeVore, "Social Behavior of Baboons and Early Man," in Sherwood Washburn, ed. *Social Life of Early Man* (Chicago: Aldine Pub. Co., 1961): 91–105.

Washburn, Sherwood. "The Study of Race," *American Anthropologist* 65 (1963): 521–31.

Wasson, David. *Yakima Herald-Republic* (January 31, 1999).

Watson, D. M. S. *Obituary Notices of Members of the Royal Society* 7:19 (November 1950): 82–93.

Weidenreich, Franz. "Giant Early Man from Java and South China," *Anthropological Papers of the American Museum of Natural History* 40:1 (New York: American Museum of Natural History, 1945).

Weidenreich, Franz. "Interpretations of the Fossil Material," in *Studies in Physical Anthropology: Early Man in the Far East*, W. W. Howells, ed. (American Association of Physical Anthropology, 1949): 149–57.

Weidenreich, Franz. *Apes, Giants and Man* (Chicago: University of Chicago, 1945).

Willey, Gordon R. and Jeremy A. Sabloff. *A History of American Archaeology* (San Francisco: WH Freeman & Co., 1974).

Williams, David. *Deformed Discourse: the Function of the Monster in Medieval Thought and Literature* (Montreal: McGill-Queens University Press, 1996).

Williams, Thomas R. *Getting Organized: A history of amateur astronomy in the United States* (Doctoral Thesis, Rice University, 2000).

Wiltsie, Gordon. "Barry Bishop, Photographer," *Mountain Research and Development* 16:3 (August 1996): 294–308.

Wolpoff, Milford and Rachel Caspari. *Race and Human Evolution: A Fatal Attraction* (New York: Simon & Schuster, 1997).

Woodley, Michael A., Darren Naish, and Hugh P. Shanahan. "How many Extant Pinniped Species Remain to be Discovered?" *History of Biology* 20:4 (December 2008): 225–35.

Wylie, Kenneth. *Bigfoot: A Personal Inquiry into the Phenomenon* (New York: Viking Press, 1980).

# Index

CPSIA information can be obtained at www.ICGtesting.com
Printed in the USA
LVOW12s1724050814

397637LV00004B/822/P